Dictionary of

MOLECULAR BIOLOGY AND BIOCHEMISTRY

Dictionary of MOLECULAR BIOLOGY AND BIOCHEMISTRY

Edited by

Er. Haojam Rocky Singh

Richa Chowdhary

***JNANADA PRAKASHAN* (P&D)**

in association with

THE GLOBAL OPEN UNIVERSITY

NAGALAND (INDIA)

Published by :

JNANADA PRAKASHAN (P&D)

4837/2, 24, Ansari Road, Daryaganj
New Delhi-110002
Phone : 011-23272047
Mobile : 9212137080
Email: jnanadabooksdelhi@yahoo.com
Website: www.jnanadabooks.com; text.ind.in

Assisted by :

TEXT BOOK PROMOTION SOCIETY OF INDIA

4837/2, 24, Ansari Road, Daryaganj
New Delhi-110002
Phone : 011-23272047
Mobile : 9212137080

First Edition: 2021

Dictionary of Molecular Biology and Biochemistry

ISBN: 978-81-7139-454-8

Paper Quality: 80 gsm sunshine
Size: Demy
Pages: 368

Typesetting by :

Vardhman Computers
New Delhi-110017

Published by Mrs. S. Chowdhary for M/s. Jnanada Prakashan (P&D) Daryaganj, Ansari Road, New Delhi-110 002 and *printed at* Balaji Offset, Navin Shahdara, Delhi-110032.

PREFACE

The study on Molecular Biology and Biochemistry is more diverse than one decade ago. The identification of the genes has rushed ahead of the biochemical characterization of their functions. Innumerable protein and nucleic acid factors have been discovered. The popularity of molecular biology is increasing with every passing day and innovative scientific research on this subject including biochemistry is redefining the evolution of man and other creatures. The students of biology are curious to know more on this subject, but there is dearth of molecular biology and biochemistry dictionary which can give answers to their inquisitive mind.

This dictionary is a sincere attempt to present an up-to-date entry of terms currently in used in molecular biology and biochemistry. It is intended to provide a convenient reference source for researchers, students and technicians. The dictionary tries to fill the vacuums that are found in other dictionaries on this subject. It tries to provide a consolidated, comprehensive and accessible list of terms and acronyms that are used in molecular biology and biochemistry.

The students of higher studies including research scholars and professors will find this dictionary very useful and informative. The editor has taken immense pain and long innings of desk work while compiling and preparing this dictionary. Suggestions for any improvement will be welcomed and will be incorporated in the next edition. Considering the need of Molecular Biology students of higher studies, this dictionary has been carefully prepared in response to their needs. There are three thousand and four hundred entries in this dictionary. Students appearing for competitive examinations will find this dictionary very helpful.

Suggestions are welcome for improving the next edition of this important publication.

A

(A+T)/(G+C) ratio: A reference to the base composition of double-stranded DNA. DNA from different sources has different ratios of the A-to-T and G-to-C base pairs, e.g. DNAs isolated from organisms that live in hot springs have a higher GC content, which takes advantage of the increased thermal stability of the GC base pair. (*see also* Chargaff's rule)

-aemia: A suffix of medical terms derived from the Greek for blood which indicates the presence of a substance in the blood.

α-imino acid: A carboxylic acid that bears a substituted amino group on its α-carbon; commonly proline, hydroxyproline, sarcosine.

AA: (=atomic absorption)

ab initio: From the beginning, as opposed to pre-existing, e.g. and *ab initio* complex is one that is assembled from its completely separated subunits, or an *ab initio* calculation is one that is constructed from theory, rather than relying in part upon empirical observations.

abortive mitosis model: The hypothesis that cells undergo a programmed death when induced to mitosis at an inappropriate phase of the cell cycle. (*see also* apoptosis)

absolute configuration: The actual, as opposed to relative to some other compound, orientation of atoms in space at an asymmetrical centre.

absorbance: $A = \varepsilon\, c \cdot l$; a quantitative definition of the dependence of the absorbance of monochromatic light, A, on the molar absorption coefficient, ε, the concentration of the chromophore, c, and the length of the light path, l.

absorption: In immunology, the neutralization of immune serum by some of the antigens against which it has been raised. Serum raised against a hapten attached to a larger protein may be exposed to the carrier protein alone in order to absorb the antibodies directed towards epitopes of the protein, leaving the serum with un-neutralized antibodies directed only towards the hapten.

absorption spectrum: The molar absorption (extinction) coefficient as a function of wavelength, usually displayed with absorbance on the ordinate and wavelength on the abscissa.

Abu shunt: An alternative pathway in bacteria and higher plants for entry of glutamate into the tricarboxylic acid cycle: glutamate→γ-aminobutyrate (Abu) → succinic acid semialdehyde → succinate.

abzyme: It means catalytic antibody (abzyme).

accelerated diffusion: It means facilitated diffusion.

acceptor: In enzyme mechanisms, a functional group of an enzyme that transiently receives a moiety of a substrate, the *donor*, before itself becoming a donor in transferring it to a second substrate, which is also an 'acceptor'. More generally in immunology, pharmacology and cell biology, an acceptor is an entity that receives an atom, ligand or structure from a 'donor'.

acceptor (3') splice site: A site in pre-mRNA that corresponds to the 3'-end of the intron and the 5'-end of the next exon. The last two bases of the intron at the acceptor splice site are AG in most cases, but AC in some instances.

acceptor stem: The formalized pattern assumed by a tRNA molecule viewed in two dimensions that shows the regions of internal complementarity that allow the polynucleotide to fold back upon itself into base-paired double helices. The stem includes the *acceptor stem* at the 3'-end, which attaches the amino acid, and the non-complementary loops or arms include the *anticodon arm*, which hybridizes with the codon of an mRNA. In three dimensions the structure can be divided into two sections at a right angle to one another: a coaxial stack that includes the acceptor stem, and the remainder of the molecule, i.e. the common arm that is shared by the two sections.

acetal: The product of reversible condensation of an aldehyde and an alcohol in which the alcoholic hydroxy group adds across the carbonyl group of the aldehyde, e.g. glucose acting as both an aldehyde and an alcohol to give the internally condensed glucopyranose; also the bond formed by this condensation. An *acetal* is the product formed by abstraction of the hydroxy group of a hemiacetal and the hydrogen of a second alcohol, e.g. a glucopyranoside or a polysaccharide.

acetate rule: The observation, first based on inspection of structures and later on experimentation, that many natural products appear to have been assembled from multiple acetate (acetyl-CoA and/or malonyl-CoA) units in head-to-tail condensations. (*see also* acetogenin; polyketide; propionate rule)

acetogenin: A compound derived from acetyl units donated by acetyl- and/or malonyl-CoA units, assembled into a non-reduced polyketide, i.e. with the carbonyl groups intact, often, then, cross-linked by aldol condensation and processed by further biochemical transformations to the final product, e.g. orsellinic acid, griseofulvin. (*see also* acetate rule; depside; polyketide)

Achilles heel cleavage: A method for cleaving of DNA at very specific sequences. The site of cleavage is able to bind a specific protein, e.g. the *lac* repressor, that already contains a restriction site or is engineered to contain one; the presence of the DNA-binding protein protects the restriction site while the rest of the DNA is enzymically methylated, thus eliminating all unprotected restriction sites. Upon removal of the binding protein, the DNA can be cleaved at the one remaining restriction site.

acid blob: A feature of DNA-binding proteins that activate yeast transcription. The only characteristic of these activating regions seems to be the high density of acidic amino acid residues.

acid protease: (= aspartate proteinase)

acinar cell: A secretory cell within an acinus.

acinus: A cluster of secretory cells surrounding a duct.

acrosome: A specialized lysosome of a spermatozoon that contains hyaluronidase, the proteinase acrosin and other hydrolytic enzymes.

ACS: Anomalously replicating consensus sequence. (*see* origin recognition complex (ORC))

action spectrum: A measure of the effectiveness of various wavelengths in promoting photosynthesis; usually a plot of photosynthetic efficiency against wavelength.

activation: In immunology, the immune response of an organism to a foreign substance; lack of a response is called tolerance (*see also* tolerance).

activation energy: The energy needed to raise the reactants, or an enzyme-substrate complex, to the transition state, where it has an equal likelihood of conversion to product or reversion to reactants. This value is commonly evaluated in an *Arrhenius plot*, ln$k$ against $1/T$, where k is the rate constant, T is the absolute temperature and the slope is E_a/R (E_a is the activation energy and R the gas constant). (*see also* Q10; reaction co-ordinate)

activator: In enzyme kinetics, a compound that increases the rate of an enzymic reaction.

active acetaldehyde: The CH_3CHOH- group when attached to the thiamin pyrophosphate prosthetic group of pyruvate decarboxylase in the pyruvate dehydrogenase complex.

active glycolaldehyde: The a-hydroxyethylidene group when attached to the thiamin pyrophosphate prosthetic group of transketolase.

active site: The binding and catalytic sites of an enzyme; more loosely, those residues of an enzyme that interact with a substrate or participate in any way in binding or catalysis.

active transport: An energy-requiring transport mechanism; one that works against a concentration gradient. (*see also* passive diffusion)

acyl-: A prefix which refers to R(CO)-, the acyl group, e.g. a fatty acyl or amino acyl group.

acyl-enzyme: An intermediate in the hydrolysis of substrates by some peptidases and esterases, e.g. by serine proteinases, in which the acyl moiety of the substrate is transiently attached to a serine hydroxy group of the enzyme.

acylium ion: (*see* a-type ion)

ad-: A prefix derived from the Latin for before which indicates something adjacent.

adaptation: The evolution of a feature or function through natural selection of incremental improvements, as contrasted with exaptation. (*see also* junk DNA)

adaptive immunity: A system of blood factors, especially the complement system, and cells, especially macrophages, which affords protection against invading micro-organisms. Unlike acquired or adaptive immunity, which is based upon recognition by antibodies, innate immunity is independent of prior exposure and depends upon pattern-recognition receptors of relatively broad specificity, particularly those on the surface of macrophages, which allow recognition of micro-organisms and opsonized particles.

adaptor hypothesis: The proposal, made before the roles of mRNA and tRNA in protein synthesis were established, that a low-molecular-mass form of RNA provides the interface between each amino acid and the template. The hypothesis assumed that the interaction of the adaptor and a polynucleotide template is much more specific than that between the amino acid and the template.

adaptor protein: One of the non-enzymic components of a multi-protein signal transduction complex. The complex is assembled upon activation of the receptor or is assembled but dormant until receptor activation. The scaffold is the framework for assembly at the cytoplasmic domain of a receptor; with the assistance of anchoring proteins it recruits kinases, phosphatases and other enzymes, and, with the assistance of adaptor proteins, other factors which will continue the signal sequence within the cell.

additivity: The principle of thermodynamics that free energy (or enthalpy of entropy) is a sum of contributions of independent components, e.g. the free energy of a protein transition is the sum of free energy changes owing to hydrogen bonding, electrostatic interactions, van der Waals packing effects, restriction of rotation about bonds, interactions with solvents, etc. The utility of this approach is undermined by the lack of a true independence of these factors, resulting in an inability to accurately evaluate their individual contributions to the overall change in free energy.

address molecule: The homing of cells or viruses to specific tissues or organs, which is facilitated by organ-selective address molecules on the target tissue surface that tightly bind to cell-specific markers on the cell or virus surface.

adeno-: A prefix derived from the Greek for gland.

adenylate energy charge: A measure, on a scale of 0-1, of the degree of phosphorylation of adenine nucleotides; ([ATP]+$^1/_3$[ADP])/([ATP]+[ADP]+[AMP]). (*see also* phosphorylation potential)

adipo-: A prefix that indicates fatty tissue, e.g. adipocyte (a fat cell).

adipose tissue: Fatty tissue.

A-DNA: A right-handed helix; a variant of the dominant B-form of DNA in solution, in which the base pairs are tilted somewhat out of the perpendicular orientation to the axis of the helix.

adrenergic: Responsive to the adrenal medullary hormone, i.e. adrenaline (epinephrine), and by extension responsive to other catecholamines. *β*-Adrenergic responses are those that result in the intracellular generation of cyclic AMP.

advanced glycation end product (AGE): The degradation product of a protein-carbohydrate addition product such as haemoglobin A_{1C}, the Schiff base formed by the condensation of an *ε*-amino group of haemoglobin with glucose.

afferent: In neuroanatomy and physiology, this describes the neuron which delivers a signal to a synapse; contrasted with efferent, which refers to the neuron which receives and propagates a signal.

affinity chromatography: The separation of soluble macromolecules by use of a stationary phase that is designed to interact specifically with, and thus retard the elution of, the desired material; e.g. a hapten attached to a resin to help isolate an immunoglobulin directed against it.

affinity cleavage: A variation of the footprinting technique in which a DNA molecule is cleaved at a site occupied by a binding protein. A metal chelator [e.g. ethylenediaminetetra-acetate (EDTA)] is covalently attached to an amino acid residue of the binding protein; Fe^{3+} is ligated to the modified binding-protein-DNA complex and reduced *in situ* to generate free radicals that cleave the DNA in the immediate vicinity. A DNA sequencing gel ladder subsequently indicates the site of binding.

affinity labelling: A technique that depends upon the tight attachment of a ligand to a binding site on a protein or cell, followed by a chemical reaction to link the ligand covalently to its binding site. (*see also* photoaffinity labelling)

affinity precipitation: A technique for purification of proteins that depends upon reversible attachment to a ligand. *A bis-ligand* (or *homofunctional ligand*) is a double-headed molecule that can attach at each end to part of a multisubunit protein, forming an insoluble lattice. In an alternative approach, a ligand attaches a protein to a water-soluble polymer that can be made insoluble by changing the conditions of pH, temperature or ionic strength.

AFM: AFP; a protein, notably present in the blood of Antarctic Ocean fish, that adsorbs to ice crystals and prevents their growth.

agarose: A highly purified agar derivative that is used as an electrophoresis and chromatography support.

AGE: The degradation product of a protein-carbohydrate addition product such as haemoglobin A_{1C}, the Schiff base formed by the condensation of an ε-amino group of haemoglobin with glucose.

agglutination: The clumping together of cells that are suspended in a fluid.

agglutinin: A compound that cross-links cells, e.g. a lectin, an antigen that reacts with sensitized cells.

aggregation: The formation of higher-molecular-mass species due to non-covalent adherence of smaller species. Especially for proteins, aggregation is a form of denaturation in which non-polar surfaces of secondary structures, e.g. those of a-helices and *β*-sheets that normally form intramolecular interactions and are buried within the interior of the protein, are allowed to interact intermolecularly and to form multimolecular forms that are sometimes insoluble. Aggregation is contrasted with *oligomerization*, the normal interaction of native, correctly folded, proteins into higher-order multimers.

aggretope: A consensus sequence that permits a protein, usually a foreign protein, to bind to a particular major histocompatibility complex protein of a T-cell.

aglycone: The non-sugar moiety of a glycoside.

agonist: A compound, often a hormone or its analogue, that binds to a receptor and elicits a response. (*see also* antagonist)

ala scan: A series of structural variants of a peptide synthesized in order to evaluate the contribution of each residue to the peptide's binding. In each variant a different residue is replaced with an alanine residue, and the affinity of the variant peptide is measured and compared with that of the original.

alarmone: A metabolite that signals a cell's response to changing environmental conditions, e.g. guanosine-3',5'-bispyrophosphate in bacteria, diadenosine-5,5'-P1,P4-tetraphosphate in blood cells.

albumin: Originally, a protein that is soluble in salt-free water and that will coagulate when heated; also the principal protein of plasma or serum. (*see also* globulin)

aldimine form: A tautomer of the Schiff base adduct of pyridoxal phosphate with an amino acid in which the pyridine ring nitrogen is protonated but uncharged and the carbon atom of the coenzyme that attaches the amino group of the amino acid has lost a proton. It is an intermediate between the *aldimine form* (in which the pyridine ring is protonated and charged and the attachment of the amino acid to the cofactor is a Schiff base) and the *ketimine form* (which can best be represented as a Schiff base formed by condensation of an a-oxo acid with pyridoxamine).

aldol condensation: The reversible formation of a bond between two carbons, one of a carbonyl group and the other adjacent to a second carbonyl group; by extension, a condensation of compounds that share that kind of chemistry, e.g. the citrate synthase reaction.

algorithm: In biotechnology, the sequence of processes in an automated program; e.g. in PCR amplification, the sequence of initial, intermediate and final heating and cooling steps and their durations.

aliquot: A representative sample of a fixed proportion. Thus 50l from a 25ml solution represents $^1/_{500}$ of the total.

alkaline Bohr effect: The decrease in the affinity of haemoglobin for oxygen that occurs when the haemoglobin solution is made more acid above pH 6. The opposite occurs below pH 6; hence the physiological phenomenon is the *alkaline Bohr effect*.

alkaloid: A nitrogen-containing natural product of a plant, often with pharmacological properties, e.g. morphine, nicotine, strychnine.

allele: One of the alternative DNA sequences that may occur at a given chromosomal locus. An individual with identical alleles at this locus on both of his homologous chromosomes is a homozygote; one with non-identical alleles is a heterozygote. Compared to the normal or wild-type, a compound heterozygote has two different mutations at the genetic locus. In a case in which one allele leads to an observable gene product and the other has no phenotype, the functional allele is said to be dominant and the non-functional allele, recessive.

allele-specific amplification: PASA; a method for genotyping of single-base mutations. Genomic DNA is used as a template in two reactions, one with a primer that makes a perfect match to the wild-type allele, another with a primer matched to the mutated allele; both reactions contain a second primer that makes a perfect match to the opposite strand of either wild-type or mutant allele. PCR will preferentially amplify fragments from an allele which is present in the genomic DNA. Characteristic results will be seen for wild-type and mutant homozygotes and for heterozygotes.

allele-specific PCR: PASA; a method for genotyping of single-base mutations. Genomic DNA is used as a template in two reactions, one with a primer that makes a perfect match to the wild-type allele, another with a primer matched to the mutated allele; both reactions contain a second primer that makes a perfect match to the opposite strand of either wild-type or mutant allele. PCR will preferentially amplify fragments from an allele which is present in the genomic DNA. Characteristic results will be seen for wild-type and mutant homozygotes and for heterozygotes.

allelic exclusion: The process by which some genes are rendered non-equivalent. The paternal or maternal allele is not expressed (allelic exclusion), or is expressed differently in different tissues.

The phenomenon is possibly directed by an imprinting box, a parent-specific polynucleotide sequence that instructs an imprinting factor, which may act by methlyation of differentially methylated regions (DMRs), and which will reversibly activate or inactivate the gene inherited from only one parent. Imprinting passes through erasure of methylation of both parental and maternal chromosomes during germ cell development, establishment of methylation, according to the sex of the gamete, and after fertilization, maintenance of the methylation status of the genome during the mitoses of embryogenesis.

all-or-none assay: A technique to measure the total amount of a functional enzyme, regardless of its efficiency or affinity for its substrate.

allostasis: The altered state caused by stress, trauma, disease, aging, etc. The cumulative burden (stated in physiological, endocrinological or psychiatric terms, depending upon the context) of the shift from homeostasis to allostasis, is the allostatic load.

allosteric effector: A compound that modifies the activity of an enzyme, or its affinity for its substrate, by binding to a site distinct from the active site; a positive effector increases the activity, a negative effector decreases the activity. (*see also* heterotropic enzyme; homotropic enzyme)

allostery: Allosteric regulation; the modification of binding or catalytic properties of a protein by binding of a regulator at a site distinct from the ligand- or substrate-binding site. Allostery typically results in sigmoid kinetics. (*see also* heterotropic enzyme; homotropic enzyme)

allotype: A classification of immunoglobulin molecules according to the antigenicity of the constant regions; a variation that is determined by a single allele. (*see also* idiotype; isotype)

alpha (α)-cell: A histological structure; usually the islet of Langerhans in the pancreas which consists of glucagon-secreting α-cells and insulin-secreting β-cells.

alpha (α)-configuration: In steroid chemistry, the orientation of substituents below the plane of the ring system, i.e. on the side opposite the angular methyl groups at C-10 and C-13, which have the *β-configuration.* Adjacent steroid rings may be *trans*-fused (the non-ring substituents are on opposite sides of the plane of the molecule) or *cis*-fused (the substituents are on the same side).

alpha (α)-helix: A secondary structure in proteins; the right-handed helical folding of a polypeptide such that amide nitrogens share their hydrogen atoms with the carbonyl oxygens of the fourth amide bonds towards the C-terminal end of the polymer. *see also* protein class

alpha (α)-isomer: In sugar chemistry, the anomer that places the hemiacetal (or hemiketal) hydroxy group on the side of the pyranose (or furanose) ring opposite the non-ring carbon atom (i.e. C-6 of glucose) that is attached to the carbon whose configuration defines the sugar as having a D- or L-configuration, i.e. C-5 of glucose; thus the mirror image of a-D-glucose is a-L-glucose. For example, for the pyranose form of a-D-glucose, the C-1 hydroxy group is on the opposite side of the ring from C-6. (*see also* Haworth projection)

alpha (α)-oxidative decarboxylation: Removal of carbon dioxide from a carboxylic acid that is facilitated by oxidation of the α-carbon (*α-oxidative decarboxylation*, e.g. the pyruvate dehydrogenase reaction) or the β-carbon (*β-oxidative decarboxylation*, e.g. the isocitrate dehydrogenase reaction).

alpha/beta (α/β)-hydrolase fold: A structure common to a group of proteins, including esterases, carboxypeptidases and dehalogenases, that have a catalytic triad of nucleophile, histidine and carboxylate positioned on a scaffolding of eight β-sheets connected by α-helices. (*see also* topology)

alpha$_1$ α_1)-cysteine proteinase inhibitor: An enzyme inhibitor that also functions as a *kininogen*; specifically H-kininogen and, in the rat, also T-kininogen. (*see also* cystatin)

alpha$_2$ (α_2)-cysteine proteinase inhibitor: (= L-kininogen; *see* kininogen)

altered-self hypothesis: The proposal that helper T-cells recognize a foreign material, e.g. a viral antigen, when it is presented on the surface of an antigen-presenting cell as a complex with the major histocompatibility complex (MHC) class II glycoprotein. This is contrasted with a less favoured explanation, the *intimacy* or *dual-recognition model*, which postulates that helper T-cells must simultaneously recognize both the foreign antigen and the separate MHC complex. By a simple extension, the altered-self hypothesis further proposes that a chemically altered MHC may also be recognized by helper T-cells.

alternative splicing: The production of more than one mRNA from a single pre-mRNA due to differences in the excision of introns and/or the use of stop codons.

Alu sequence: A small interespersed repeat element (SINE); a repetitive sequence in human DNA found in some introns, and characterized by Alu restriction sites; the locus of some homologous recombinations. Comprising 3-6% of the human genome, the sequences are about 300 bp long and usually feature a head-to-tail tandem repeat.

Alu-PCR: A method for amplification of human DNA sequences in hybrid somatic cells, using primers directed at species-specific repetitive sequences: a short interspersed repeat element (e.g. *Alu*) sequence, a long interspersed repeat sequence or a combination of both.

amber mutation: A nonsense mutation; the formation of a non-functional protein due to the premature appearance in mRNA of the terminator codon UAG. (*see also* ocher mutation; opal mutation)

ambidexteran: An aggregate of glycosaminoglycans that, unless prevented by steric hindrance or over- or under-sulphation, presents one or both surfaces of their tape-like shape to the complementary surface of another; this results in linear polymers if aggregation is on only one surface, or sheets if aggregation can occur on both surfaces.

ambo: A prefix denoting a (not necessarily equal) mixture of D- and L-isomers, e.g. in an amino acid sequence, Ala-*ambo*-Glu-Gly=Ala-Glu-Gly+Ala-D-Glu-Gly.

ames test (Salmonella test): A test for mutagenicity and carcinogenicity which uses specially constructed bacterial strains, by screening the effects of test compounds for their ability to produce reverse mutations that will restore the ability of the bacteria to grow in the absence of an essential metabolite.

aminimide: A peptide analogue in which an amino nitrogen substitutes the α-amino group of an amino acid residue; this forms a stable, soluble product.

amino-: A prefix which refers to $-NH_2$, the amino group.

amino acid: Usually an α-amino acid, in which a carboxy and an amino (or imino) group are attached to the α-carbon; triple- and single-letter codes are shown:

alanine	Ala	A
arginine	Arg	R
asparagine	Asn	N
aspartic acid	Asp	D
cysteine	Cys	C
glutamic acid	Glu	E
glutamine	Gln	Q
glycine	Gly	G
histidine	His	H
isoleucine	Ile	I
leucine	Leu	L
lysine	Lys	K
methionine	Met	M
phenylalanine	Phe	F
proline	Pro	P
serine	Ser	S
threonine	Thr	T
tryptophan	Trp	W
tyrosine	Tyr	Y
valine	Val	V
aspartic acid or asparagine	Asx	B
any residue that upon acid hydrolysis yields glutamic acid (Glu, Gln, Glp, Gla)	Glx	Z
4-carboxyglutamic acid	Gla	
hydroxyproline	Hyp	
pyroglutamic acid	Pgl/Glu/Glp	pE
unknown or unspecified	Xaa	

amino acyl site: The part of a ribosome that binds one amino acyl-tRNA where it will accept the peptidyl group held at the peptide site in the form of its tRNA ester. (*see also* peptide site)

ammonotelic: Descriptive of an organism which excretes ammonia as an end product of nitrogen metabolism, such excretion being ammonotely. (*see also* ureotelic; uricotelic)

ammonotely: *see* ammonotelic

amniocentesis: A procedure to obtain fetal cells during pregnancy by puncturing the womb with a needle and removing some fluid that surrounds the fetus. The procedure is used, in part, to obtain fetal cells from which DNA can be isolated for prenatal genetic analysis.

amphibolic pathway: A metabolic pathway that participates in both anabolic and catabolic pathways, e.g. the tricarboxylic acid cycle.

amphipathic helix: A protein structure that serves in part as an interface between polar and non-polar phases; an a-helix that displays non-polar residues on one side and polar residues on the other (e.g. in many globular proteins).

amphipathic: Having both polar and non-polar groups, e.g. a detergent.

amphitropic: Having an affinity for both lipid and aqueous environments, e.g. a membrane-associated protein that has domains that are embedded in the membrane and others that extend into the cytoplasm or the extracellular space.

ampholyte: A molecule with both an acidic and a basic group, e.g. an amino acid. (*see also* polyampholyte)

amphoteric: Having both acidic and basic groups, e.g. an amino acid.

amplicon: A cloned, amplified (by PCR), DNA sequence. (*see also* polymerase chain reaction (PCR); representational difference analysis (RDA))

amplification of refractory mutation system: (*see* PCR amplification of specific alleles)

amplification: The causation of a quantitatively major biochemical or physiological event by a quantitatively minor initiator, e.g. the blood clotting cascade, ion gating; also, using the polymerase chain reaction, the increase by orders of magnitude of copies of a DNA template.

amplimer: A PCR primer. (*see also* polymerase chain reaction (PCR))

amyloplast: An organelle of plants that synthesizes and stores starch.

ana-: A prefix derived from the Greek for up and which refers to synthetic processes.

anabolism: Those energy-requiring metabolic pathways that result in synthesis of macromolecules and their building blocks, e.g. gluconeogenesis, fatty acid synthesis. (*see also* amphibolic pathway; catabolism)

anabolite: A metabolite built up from more simple compounds. (*see also* catabolite)

anaerobic glycolysis: One of the central pathways of metabolism in most eukaryotes and many other cells; the sequence of enzymic reactions that converts glucose into lactic acid (*anaerobic glycolysis*) or into pyruvate (*aerobic glycolysis*).

analyte: Material evaluated in an assay.

analytical ultracentrifugation: A technique of very-high-speed centrifugation that sediments soluble macromolecules and characterizes them according to their rate of sedimentation (*sedimentation-velocity ultracentrifugation*) or the extent of their sedimentation (*equilibrium sedimentation ultracentrifugation*). (*see also* density-gradient centrifugation)

anaplerotic pathway: Metabolic reactions that replenish the pools of intermediates of the tricarboxylic acid cycle. These pools may become depleted, as they also serve as precursors for amino acid synthesis, gluconeogenesis and other anabolic reactions.

anchored PCR: A variation of the PCR technique, similar to ligation-mediated PCR, that is applied to double-stranded DNA fragments for which the sequence at only one end of the gene is known. The technique allows amplification of a complete sequence of a gene when only the N-terminal sequence of a protein is known. A short polynucleotide of known sequence is ligated to the 3'-ends of the double-stranded DNA so that a primer complementary to it can be added, along with the primer determined from the partial protein sequence. (*see also* ligand-mediated PCR (LMPCR))

anchored reference marker: One of many genes that are used in mapping of chromosomes. Having a well defined locus within a chromosome, it is used to test for genetic linkage in order to localize new traits; also, in comparative genetics, to test for conservation of linkages which may be of functional significance.

anchoring protein: One of the non-enzymic components of a multi-protein signal transduction complex. The complex is assembled upon activation of the receptor or is assembled but dormant until receptor activation. The scaffold is the framework for assembly at the cytoplasmic domain of a receptor; with the assistance of anchoring proteins it recruits kinases, phosphatases and other enzymes, and, with the assistance of adaptor proteins, other factors which will continue the signal sequence within the cell.

androgen: A compound, usually a steroid, that supports the development of male secondary sex characteristics, e.g. testosterone.

anergy: In cell biology, unresponsiveness or inactivation, e.g. clonal anergy, the inactivation of a lymphocyte, as opposed to clonal proliferation or clonal deletion.

aneuploidy: (*see* euploidy)

anfinsen cage: A proposed model for chaperone-assisted protein folding in which a chaperone, GroEL for instance, provides an environment in which a denatured protein can unfold and correctly refold, undisturbed by the possibility of aggregation with other unfolded molecules.

angio-: A prefix derived from the Greek for vessel which refers to blood vessels.

anhydro-: A prefix which indicates a structure from which water, or the elements of water, i.e. a hydrogen atom and a hydroxyl group, has been removed.

anion: A ion attracted to the anode of an electrolytic cell, which being positively charged, attracts anions, which are, therefore, negatively charged. Similarly, a cation, which is attracted to the negatively charged cathode, is positively charged.

anisotropic: Descriptive of a physical property that varies with the angle of observation. (*see also* isotropic)

anisotropy: A measure of the mobility of a fluorophore: $A=(I|-I\perp)/(I|+2I\perp)$, where I is the intensity of emission and $|$ and $\perp$ indicate polarization parallel and perpendicular respectively to the exciting light. A mobile fluorophore is able to reorientate itself within its fluorescence lifetime and therefore emits unpolarized light; an immobile fluorophore does not reorientate itself, and hence it emits light polarized in the plane of the excited light. (*see also* polarization)

annealing (hybridization): The time- and temperature-dependent process by which two complementary single-stranded polynucleotides associate to form a double helix.

anode: A ion attracted to the anode of an electrolytic cell, which being positively charged, attracts anions, which are, therefore, negatively charged. Similarly, a cation, which is attracted to the negatively charged cathode, is positively charged.

anoikis: Apoptosis that is associated with detachment of cells from their matrix or their attachment to the 'wrong' molecules.

anomalous scattering (resonance scattering): A property of diffracted radiation that is useful in X-ray crystallographic analysis for identification of diffracting centres. The otherwise symmetrical diffraction pattern is subtly perturbed when incident energy of an appropriate wavelength interacts with a diffraction centre (in practice, a sulphur or heavy metal atom) that changes the phase of the diffracted radiation in a manner that is characteristic of that atom.

anomalously replicating sequence: (*see* origin recognition complex (ORC))

anomer: One of two possible compounds that arise when the open-chain form of a sugar condenses via a hemiacetal or hemiketal bond and produces a new asymmetrical centre. (*see also* alpha (α)-isomer; beta (β)-isomer)

antagonist: A compound, often an analogue of a hormone, that binds to a receptor but elicits no response. (*see also* agonist)

anterograde: (*see* retrograde)

anthrogenic: Descriptive of a cell in which all the chromosomes are of paternal origin; as opposed to *gynogenic*, descriptive of a cell in which all the chromosomes are of maternal origin.

anthrone reaction: A colorimetric method for estimation of sugars that involves treatment with sulphuric acid and anthrone.

anti conformation: In nucleic acid chemistry, the orientation about the glycosidic bond of a nucleoside or nucleotide that places the base away from the sugar moiety; contrasted with the *syn conformation*, in which the base and sugar are oriented towards each other. (*see also* Z-DNA)

antibiotic: A natural, synthetic or semi-synthetic product, especially a pharmaceutical, that inhibits bacterial growth.

antibody: An immunoglobulin molecule that reacts specifically with another (usually foreign) molecule, the *antigen*.

antichimaeric assistance: (*see* entropy effect)

anticodon arm: (*see* cloverleaf)

anticodon: The three-nucleotide sequence of a tRNA molecule that is complementary to a triplet of mRNA (the *codon*) which specifies a certain amino acid.

anti-ergotypic: Descriptive of killer T-cells that recognize and respond to a cell- surface marker on a T-cell that is actively secreting immunoglobulin; contrasted with *anti-idiotypic*, descriptive of killer T-cells that are more restricted in that they recognize and respond to a cell-surface marker on T-cells that produce a specific immunoglobulin idiotype.

antifolate: An antimetabolite that blocks the action of tetrahydrofolic acid-dependent reactions, usually by inhibition of folic acid reductase.

antifreeze protein: AFP; a protein, notably present in the blood of Antarctic Ocean fish, that adsorbs to ice crystals and prevents their growth.

antigen (immunogen): A substance that causes production of an antibody directed against itself.

antigen presentation: The appearance on the surface of a cell of a foreign protein, e.g. one that arises from viral infection, in a complex with a class I major histocompatibility complex protein. It is this complex that is recognized by killer T-cells.

antigenic determinant: (= epitope)

antigenic mimicry: The phenomenon of two proteins having very similar epitopes, so that antibodies raised against one, for instance a foreign protein, will also react with a second, for instance an endogenous protein.

antigenized antibody: A molecularly engineered antibody that incorporates epitopes of a non-antibody antigen in the complementarity determining regions of the heavy- and light-chain V domains.

anti-idiotypic: (*see* anti-ergotypic)

anti-Lepore haemoglobin: (*see* Lepore haemoglobin)

antimetabolite: An inhibitor of a key enzyme in metabolism, used to suppress the activity of the cell; often used in chemotherapy.

anti-mutator DNA polymerase: A polymerase with a higher than usual degree of fidelity in proofreading.

anti-oncogene (tumour suppressor gene): A gene that normally functions to regulate cell proliferation by suppression of the function of an oncogene; one cause of cancer is the loss or damage of anti-oncogenes.

antiparallel: In protein chemistry, the orientation of extended polypeptide chains that interact in a pleated sheet structure, one chain in an N- to C-terminal direction and the other in a C- to N-terminal direction; in nucleic acid chemistry, the orientation of the two polynucleotide chains of a double helix, one that runs in a 3' to 5' direction and the other in a 5' to 3' direction.

antiport: A transport mechanism that simultaneously drives two different compounds or ions in opposite directions across a membrane. (*see also* mobile barrier; mobile carrier; uniport)

antisense drug: (= code blocker (antisense drug))

antisense: Descriptive of an endogenous or semi-synthetically produced oligoribonucleotide complementary to mRNA and capable of base-pairing and annealing with mRNA to prevent translation; or of an oligodeoxyribonucleotide capable of binding to the major groove of polypurine-polypyrimidine sequences of DNA by Hoogsteen base pairing to silence a gene. Also used to describe one of the two strands of double-stranded DNA, usually that which has the same sequence as the mRNA, i.e. the non-transcribed strand. However, there is not universal agreement on this convention, and a preferred designation is *coding strand* for the strand whose sequence matches that of the mRNA, and *non-coding strand* for the complementary strand (i.e. the transcription template, or transcribed strand).

anti-terminator: A bacteriophage protein that prevents the normal termination of transcription, e.g. the N protein that binds to nut (N utilization) sites, thus countering the action of the rho protein.

AP site: A position in a double-stranded DNA sequence that is missing an A or G base (apurinic) or a C or T base (apyrimidinic).

apical (luminal): Descriptive of the free border of an epithelial cell, where it is in contact with vascular space. (*see also* basolateral)

apicomplexan: Descriptive of a class of protozoan parasites that pass part of their life cycles in mammalian cells, e.g. the malaria pathogen *Plasmodium falciparum*, and which contain in their apices organelles that assist penetration of host cells. Also characteristic are the apicoplast, an organelle which contains maternally-inherited circular DNA, possibly of green algal evolutionary origin, and enzymes of the C6C3 metabolic pathway.

apicoplast: (*see* apicomplexan)

apo-: A prefix derived from the Greek for separate which indicates a structure that is incomplete because of its separation from a natural or common ligand.

apocrine: Descriptive of a secretion mechanism in which vesicles that contain the product burst through the cell membrane and are released along with some of the cytoplasm and plasma membrane. (*see also* exocytosis; holocrine)

apolar water-accessible surface area: The surface of a protein that is internalized in native proteins and that becomes available for interaction with water during denaturation.

apoprotein: A protein stripped of any prosthetic group or metal ion normally associated with it. (*see also* holoprotein)

apoptogen: An agent that causes apoptosis.

apoptosis: A morphologically characterized process of programmed cell-death, initiated by various physiological or pathological causes (e.g. cell turnover, hormone-induced atrophy, cell-mediated immune cytolysis, tumour regression), that is characterized by shrinkage of the nucleus and cytoplasm, cell fragmentation and phagocytosis. Apoptosis is contrasted with *necrosis*, which is a random pathological process initiated by irreversible cell damage. Apoptosis is controlled by extracellular signals or the removal of extracellular suppressors of cell death. (*see also* death gene; *see also* caspase)

apparent K_m: The Michaelis constant as observed under conditions (e.g. the presence of a competitive inhibitor) that would hinder the determination of its true value; in the case of a two-substrate enzyme, the Michaelis constant measured under the particular conditions of a defined concentration of the invariant substrate.

approximation: (*see* entropy effect)

aprotinin: Bovine pancreatic proteinase inhibitor.

aptamer: A ssDNA fragment, typically one selected from a library of random synthetic sequences for affinity to a protein, by SELEX cycles (cyclic amplification and selection of targets)

apurinic DNA: A polynucleotide that has lost one or more purine bases due to the lability in acid of the glycosidic bond to purines. (*see* apyridinic DNA)

apyridinic DNA: A polynucleotide that has lost one or more pyrimidine bases. (*see* apurinic DNA)

arbitrary primer: (*see* primer).

archaea: Common name of a member of the *Eucarya*, one of the three domains of living organisms that are classified according to rRNA sequence homologies; they are further characterized by cells containing nuclei and membranes composed primarily of diacylglycerol derivatives. The other two domains are the

Eubacteria (the bacteria), which have no nuclei (i.e. prokaryotes) and also have diacylglycerol derivative membrane components, and the *Archaea* (archaebacteria), which are also prokaryotes; they have a membrane composed of isoprenyl glycerol diether and tetraether lipids. Kingdoms within the *Eucarya* include plants, animals and fungi; kingdoms of the *Archae* are euryarchaeotes (or euryotes), which include methanogenic, halophilic, sulphur-reducing and some thermophilic organisms, and the crenarchaeotes (or crenotes), which include other sulphur-dependent organisms, thermoacidophiles and extreme thermophiles (eocytes).

archaebacteria: (*see* eukaryote)

area detector: An 'electronic film'; a device for measurement of diffracted radiation, e.g. in X-ray crystallographic analysis; detector devices are densely arrayed over a screen and report to a computer the intensity of incident light and its position on the screen. (*see also* image plate; multiwire data collector)

arginine fork: A specific kind of interaction between an RNA and an RNA-binding protein in which the two equivalent guanidinium nitrogens of an arginine residue interact with adjacent phosphates of a non-double-stranded region of the polynucleotide.

ARMS: Amplification of refractory mutation system. (*see* PCR amplification of specific alleles)

array: (*see* chip (oligonucleotide array))

arrest: (*see* halt)

ARS: Anomalously replicating sequence. (*see* origin recognition complex (ORC))

aryl-: A prefix which refers to an aromatic group.

ASA: Allele-specific amplification (*see* PCR amplification of specific alleles) and apolar water-accessible surface area

ascus: A spore-like form through which yeast cells pass during their life cycle; an outer wall surrounds a diploid cell that undergoes meiosis to form four haploid cells (ascospores) which eventually rupture to yield two diploid daughters.

AS-ODN: Antisense oligodeoxyribonucleotide. (*see* antisense)

aspartate proteinase: A type of peptidase that has at its active site two aspartate residues. (*see also* cysteine proteinase (thiol proteinase); metalloproteinase; serine proteinase)

association analysis: (*see* linkage disequilibrium)

association constant (K_a): Reciprocal of K_d (*see also* pKa)

asymmetrical PCR: A protocol for generation of single-stranded DNA. Unequal amounts of primers are used, so that the first PCR cycles generate equal amounts of each strand of the template but later cycles, which have no more of one of the primers, create only one new strand.

asymmetrical reaction: The unequal handling of like groups in a prochiral compound. (*see also* meso-carbon; Ogston hypothesis)

a-type ion: In mass spectrometry, an N-terminal fragment of a polypeptide produced by cleavage of the bond between an α-carbon and its neighbouring α-carboxy carbon. A *b-type ion* is an α-N-terminal acylium ion which results from cleavage of an amide bond, and *y-type ions* are C-terminal iminium ions from such a cleavage. (*see also* mass spectrometry (MS))

AT queue: A model that accounts for chromosomal banding. The microscopic appearance of chromosomes during mitosis (metaphase) displays *chromomeres* and *interchromomeres*, i.e. *G bands* (Giemsa stain) and *R bands* (Reverse) respectively; the former are characterized as AT-rich and gene-poor and by late replication, and the latter as GC- and gene-rich and by early replication. The DNA is presumably organized into wide spring-like coils from which 100kb G loops extend parallel to the chromosome axis (G bands), extended DNA sequences from which larger loops extend perpendicular to the chromosome axis (R loops), and AT-rich sequences called *matrix-attachment* or *scaffold-associated regions* that anchor the G and R loop regions and create the characteristically stained bands.

AT-AC intron: One of a minority of introns which have at their 5'- and 3'-termini, respectively, the AT and AC sequences. They are excised by a mechanism similar to that for the more common GU-AG introns.

atomic absorption analysis: A technique for the quantification of small amounts of a metal in solution. Monochromatic light, often generated by excitation of the element in question, passes through a flame in which a test solution is vaporized, and the non-absorbed light, corrected for the excited emission of the metal in the flame, is detected by a photomultiplier.

atomic absorption spectrometry (AA): A method for quantifying elements, usually metals, in biological samples. The method depends upon the absorption of energy by atoms as they are excited in their electronic ground state. The choice of wavelength depends therefore, on the element to be evaluated and the most convenient energy level at which it can be excited. The sample is vapourized in a flame or electric furnace, and monochromatic light of the desired wavelength is shone through the flame or cloud; absorption of the light is proportional to the number of atoms in the observed volume. In a related technique, flame emission photometry, the small numbers of excited atoms in the flame or cloud are detected as they fall to the ground state and emit light at a wavelength which is characteristic of the element and at an intensity which is proportional to the number of atoms in the observed volume. This proportionality is lost when the concentration of atoms is sufficiently high to allow self-absorption, a phenomenon akin to the inner filter effect of fluorescence spectroscopy, in which emission from the centre of the flame or cloud is absorbed by atoms on its periphery. A standard curve, constructed with known concentrations of the element, allows quantification of experimental samples.

atomic force microscopy (AFM): Also known as scanning force microscopy; a method for mapping the surface of microscopic (e.g. cellular) and even submicroscopic (e.g. macromolecular) surfaces. As a sharp tip passes over the surface of an object, contact pressure is kept at a minimal constant value by an electronic feedback loop; the whole deflection of the cantilevered tip is measured, for example by deflection of a reflected laser beam.

ATPase: An enzyme that hydrolyses ATP; usually the partial activity of an enzyme, or system of enzymes, that uses the energy made available by the hydrolysis of ATP to drive an energetically unfavourable process, e.g. the Na^+/K^+-ATPase of cell membranes.

atrophy: The wasting away of an organ and/or its capabilities.

attenuation: The response of the synthesis of bacterial mRNA to the nutritional state of the organism, e.g. the decrease in transcription of the *trp* operon in the presence of tryptophan, which is due to the incomplete transcription of a leader mRNA sequence coded for by the attenuator sequence.

attenuator: A polynucleotide sequence that occurs between an operon and its closest structural gene. (*see also* attenuation)

atto-: A prefix which denotes 10^{-18}; e.g. attomolar, 10^{-18} M.

auto-: A prefix derived from the Greek for self which indicates independence or action of an entity upon itself.

autocatalysis: The activation of a proenzyme preparation by that fraction of it that has already been activated.

autocrine: Descriptive of a secretion that binds to receptors on the surface of the same cell that produces it. (*see also* endocrine; exocrine; paracrine)

autogenous regulation: A phenomenon in which a gene for a single protein is regulated by its own promoter and operator, and constitutes a one-protein operon.

autoimmune disease: A failure of tolerance; the reaction of an individual's immune system towards some of the individual's own proteins as if they were foreign proteins, e.g. myasthenia gravis, rheumatoid arthritis.

autonomic nervous system: A functional division of the peripheral nervous system that consists of those pathways that are under involuntary control, e.g those that regulate the gastrointestinal tract and glandular function.

autophagocytosis: The response of a cell to stress in which it forms an *autophagosome*, an organelle that contains parts of the cell's own cytoplasm. Later this will fuse with a lysosome to form an *autolysosome*; all such inclusions that derive from autophagocytosis are termed *autophagic vesicles*.

autophagy: The action of a lysosome to digest materials from its own cell. (*see also* heterophagy)

autoradiography: A technique for visualization of radioactivity in histological preparations, paper chromatograms or slab gels from electrophoresis by overlaying the surface with X-ray film and allowing the radiation to form an image on the film.

autosomal recessive: Descriptive of a non-sex-linked genetic trait that must be inherited from both parents to be expressed at the phenotypic level.

autosome: A chromosome that is neither of the sex chromosomes, e.g. in humans, a chromosome that is neither an X or Y.

autotroph: A cell that can sustain itself on non-organic nutrients, e.g. a photosynthetic cell.

autotrophic: In microbiology, descriptive of organisms able to grow on inorganic material alone. (*see also* heterotrophic; mixotrophic)

auxin: A type of plant hormone that affects cell size, e.g. indoleacetic acid.

axial ratio: A measure of the asymmetry of a macromolecule, assumed to be an ellipsoid, given by the ratio of the major axis to the minor axis; it is evaluated from physical properties, e.g. hydrodynamic behaviour, light scattering.

axial: Descriptive of the orientation of a substituent on a six-membered ring that is perpendicular to the plane of the ring; the opposite of *equatorial*, which is the orientation substantially within the plane of the ring. (*see also* chair form)

axon: A long projection of a neuron through which it communicates with other cells.

axoneme: The fundamental structural unit of eukaryotic flagella and cilia; it is composed of nine microtubular doublets (in cross-section a microtubule with another fused to it) surrounding two microtubule singlets. (*see also* triplet)

axoplasm: The cytoplasm of an axon.

B

-blast: A suffix derived from the Greek for sprout which indicates an embryonic or stem cell.

BAC: Abbreviation for bacterial artificial chromosome.

backbone: In chemistry, the longest continuous chain of atoms bonded to each other, exclusive of all others, that comprise a polymer.

backbone cyclization: An approach to design of polypeptide-like drugs, in which the C- or N-termini or side chains are linked to each other (end-to-end, side chain-to-side chain, or end-to-side chain) in order to impose conformational restraints and thus limit proteolytic degradation and enhance binding to a target site.

bacteria: (*see* eukaryote)

bacterial artificial chromosome (BAC): A cloning vector that can accept up to 350-kb fragments for cloning and sequencing of fragments of the human genome. Phage artificial chromosomes are similarly used. (*see also* cosmid; yeast artificial chromosome (YAC))

bacteriophage: A virus that infects bacteria. Many phage have proved useful in the study of molecular biology and as vectors for the transfer of genetic information between cells. *Lytic phage*, e.g. the T series phage that infect *E. coli* (*coliphages*), invariably lyse a cell following infection; *temperate phage*, e.g. lambda bacteriophage, can also undergo a lytic cycle or can enter a lysogenic cycle, in which the phage DNA is incorporated into that of the host, awaiting a signal that initiates events leading to replication of the virus and lysis of the host cell.

bait region: A sequence of an α_2-macroglobulin molecule, and of homologous proteins, that contains scissile peptide bonds for those proteinases that it inhibits. (*see also* hook region)

bait: *see* two-hybrid system.

BAL: Abbreviation for British anti-lewisite.

balanced polymorphism: The result of selective pressures for and against a deleterious mutation that permits it to persist in a population. An example is the stable presence in Africa of the sickle cell gene due to the protection against malaria enjoyed by heterozygotes, i.e. those with sickle cell trait.

band competition assay: (= gel retardation assay; *see* gel shift assay (electrophoretic mobility shift assay; EMSA))

band compression: An electrophoresis artifact in which DNA fragments differing in length by only one nucleotide are unresolved, sometimes observed through a series of consecutive guanine nucleotides.

barophile: (*see* extremophile)

basal metabolic rate: (= resting metabolic rate (RMR))

base: In nucleic acid chemistry, one of the nitrogenous compounds, purines or pyrimidines, that are incorporated into nucleosides, nucleotides or nucleic acids. The most common bases are adenine, cytosine, guanine thymine and uracil. Single-letter abbreviations for the bases are:

adenine	A
cytosine	C
guanine	G
any except guanine	H
hypoxanthine (the base of inosine)	I
a purine	R
thymine	T
uracil	U
a pyrimidine	Y

base equivalency rule: (= Chargaff's rule)

base flipping: The distortion of a double-stranded DNA structure that disrupts a base pair and redirects one nucleoside of the pair outwards, where it can interact with a DNA-modifying enzyme such as a methyltransferase.

base pair: In a nucleic acid double helix, a purine and a pyrimidine on different strands that interact by hydrogen bonding, most commonly a GC or AT pair.

base roll: Variation in orientation of bases in a DNA double helix that permits some tilting of the bases. (*see also* A-DNA; Z-DNA)

basement membrane: An extracellular network of fibres and glycoconjugates that underlies and strengthens some tissues; an interface between these tissues and the connective tissue that surrounds them.

basolateral: Descriptive of the border of an epithelial cell that is attached to the basement membrane. (*see also* apical (luminal))

basophil: A polymorphonuclear leucocyte containing granules that react with a histological stain for basic substances.

batho-: A prefix derived from the Greek for depth which indicates a lowered value.

bathochromic shift: A shift to longer wavelengths.

B-cell (B-lymphocyte): A lymphocyte that can be activated to proliferate and to synthesize and secrete a specific immunoglobulin G by binding of the antigen that is recognized by the immunoglobulin G receptor on the cell.

BDA: Abbreviation for boomerang DNA amplification.

B-DNA: A right-handed helix; the dominant conformational variant of DNA in solution, in which the base pairs are stacked nearly perpendicular to the axis of the helix. (*see also* A-DNA; Z-DNA; P-DNA)

Beer's law: A quantitative treatment of the absorption of monochromatic light by a solution. The equation $A=\log(I_0/I)$ defines A (the absorbance, which is proportional to the concentration of the chromophore), I_0 (the intensity of the incident light) and I (the intensity of light having passed through 1cm of the solution). *T*, the transmission, is defined as *A*=-log*T*. (*see also* Beer-Lambert equation)

Beer-Lambert equation: $A=\varepsilon\ c\cdot l$; a quantitative definition of the dependence of the absorbance of monochromatic light, *A*, on the molar absorption coefficient, *ε*, the concentration of the chromophore, *c*, and the length of the light path, *l*.

Bence Jones protein: The immunoglobulin light chains that are synthesized in large amounts and are secreted into the urine by multiple myeloma patients.

Benedict's solution: (*see* reducing sugar)

beta (*β*)-amino acid: A zwitterionic compound with carboxyl and amino groups on adjacent carbons: H_2N-CHR-CHR'-COOH.

beta (*β*)-barrel: A form of supersecondary structure: a structure within a globular protein in which six or more extended polypeptide strands (*β*-pleated sheets) are arrayed cylindrically to form the staves of a barrel.

beta (*β*)-bend: (= reverse turn).

beta (*β*)-cell: (*see* islet)

beta (*β*)-configuration: (*see* alpha (*α*)-configuration)

beta (β)-isomer: In sugar chemistry, the anomer that places the hemiacetal (or hemiketal) hydroxy group on the same side of the pyranose (or furanose) ring as the non-ring carbon atom. (*see also* alpha (*α*)-isomer)

beta (*β*)-oxidation: The series of enzymic reactions that oxidizes fatty acyl-CoA esters and shortens them by removal of the C-terminal two carbon atoms as acetyl-CoA. More narrowly, it is the oxidation of a compound, such as a fatty acid, at the *β*-carbon.

beta (*β*)-oxidative decarboxylation: Removal of carbon dioxide from a carboxylic acid that is facilitated by oxidation of the *α*-carbon (*α-oxidative decarboxylation*, e.g. the pyruvate dehydrogenase reaction) or the *β*-carbon (*β-oxidative decarboxylation*, e.g. the isocitrate dehydrogenase reaction).

beta (*β*)-peptide: An oligomer of *β*-amino acids

beta (β)-pleated sheet: A form of secondary structure of a protein in which the amide hydrogens of a peptide bond of one extended polypeptide sequence are shared with the carbonyl oxygens of a peptide bond on a second polypeptide sequence. A sheet that often consists of three or more polypeptide sequences is said to be parallel (i.e. both adjacent strands run in the same direction; N- to C-terminal) or antiparallel. *see also* protein class.

beta (β)-ribbon: A DNA-recognition motif; a segment of antiparallel β-structure that can fit into the major groove of a double-stranded polynucleotide and bind to it due to very specific interactions between side chains of the protein and phosphate residues and edges of base pairs of the polynucleotide.

beta (β)-structure: (= beta (β)-pleated sheet)

beta (β)-turn: (= reverse turn)

beta-beta (β-β)-hairpin: A form of secondary structure in proteins in which an extended polypeptide chain turns sharply to fold back upon itself and forms an antiparallel β-sheet.

bi: (*see* enzyme mechanism)

bi-: A form of the prefix bis-. *see* di-

BIA: Abbreviation for biomolecular interaction analysis.

BIAcore: An optical instrument for detection of binding partners. A reflective surface is coated with a one of the pair, and is then exposed to a test solution. If binding occurs, the refractive index of the surface is changed and the change is detected by the optical system

bi-antennary: (*see* tetra-antennary)

bicistronic: Having two cistrons in tandem, e.g. the A and B subunits for the *Escherichia coli* enterotoxin

bidirectional PCR amplification of specific alleles (bi-PASA): A method for determination of zygosity. Two sets of primers are used. Of the inner set, primer A is complementary to a point mutation on the coding strand and primer B is complementary to the wild-type non-coding strand. The outer set of primers, P and Q, flank the mutation site on the coding and non-coding strands, respectively. P and Q are chosen so that the amplicons PB and AQ will have characteristic lengths. The pattern of PCR products is informative of the zygosity of the tested genomic DNA: all samples will generate PQ; the heterozygote and the homozygous wild-type will generate PB and the heterozygote and homozygous mutant will generate AQ. A and B are designed to mismatch the unintended allele at their 3' ends and they possess G+C-rich tails. The efficiency of replication is affected by template transfer: as replication of an amplicon proceeds, it replaces genomic DNA as the template. In bi-PASA, as PQ accumulates, it also can serve as a template for amplification of the shorter fragments. This is contrasted with self amplification, which occurs when the shorter templates replicate only themselves. Self amplification is favored by the G+C-rich tails and by annealing conditions that discourage all but the stronger G-C bonds. Basically similar to bi-PASA is tetra-primer PCR, in which the internal set of primers are mismatched in the middle of their sequences, and they lack G+C-rich tails. This variation requires high- and low-stringency annealing conditions to generate appropriate amounts of the three potential amplicons, whereas bi-PASA uses a constant annealing temperature.

bidirectional replication: Synthesis of DNA that is effected by two replication forks that travel away from a single origin of replication.

bile acid: One of the products of cholesterol hydroxylation and side-chain oxidation to the level of a carboxylic acid. The carboxylate is often conjugated through an amide bond to a glycine or a cysteic acid. Excreted into the small intestine from the gall bladder, bile acids act as detergents, and aid lipid absorption.

bile pigment: One of the highly coloured products of haem degradation.

bilin: A tetrapyrrole pigment chemically related to the bile pigments, e.g. phycobilin, a bilin of red or blue-green algae.

bimolecular sheet: (= lipid bilayer (bimolecular sheet))

binary system: In transgenic research, an approach to control expression of one transgene by a second, each initially established in its own pedigree. By crossing the two lines, doubly transgenic animals are created in which the control may become operational. The gene of interest is regulated by an exogenous ligand acting either as a positive regulator that binds to a repressor or as a negative regulator that binds to a transactivator. The repressor or transactivator are products of the second transgene.

binding protein: A circulating protein that carries its ligand from one site in the body to another, e.g. thyroxine-binding protein; also any protein specialized for binding a ligand, e.g. a calcium-binding protein.

binding site: That region of the surface of an enzyme (or receptor, or binding or transport protein) that holds the substrate or product (or other ligand).

bioactive: Descriptive of the chemical, stereochemical or conformational form of a substance in which it performs its biological function.

bioavailability: The ease of access a drug or nutrient may have to its site of biological action; e.g. the ability of an antisense oligonucleotide designed to target a neuroreceptor to cross the blood-brain barrier, the ability of vitamin B12 to be absorbed by the digestive system.

biochemical genomics: An experimental approach to identification of the gene responsible for a given enzymic activity. Each open reading frame (ORF) of an organism, yeast for instance, is fused to the cDNA of glutathione S-transferase (GST) in a plasmid and then introduced into yeast. Yeast strains containing these plasmids are assayed in groups for the desired activity until the one strain with the activity is identified. (GST facilitates purification of the GST-ORF fusion protein by affinity chromatography.) Once the strain containing the activity is known, the ORF in the plasmid, which is responsible for the activity, is easily determined.

bio-imprinting: The induction of a new catalytic specificity in an enzyme in an organic solvent. An enzyme and a weakly binding non-substrate/non-product (e.g. chymotrypsin and *N*-acetyl-D-tryptophan) are suddenly precipitated together from an aqueous solution by addition of a miscible organic solvent. The complex reverts to the native conformation and behaviour when redissolved in water, but if redissolved in a suitable organic solvent it will be found to possess a new enzymic activity, e.g. for synthesis of *N*-acetyl-D-tryptophan ethyl ester from *N*-acetyl-D-tryptophan and ethanol. (*see also* molecular imprinting)

bioinformatics: The management, manipulation and use of DNA nucleotide sequences, and consequent protein amino acid sequences, that are known from sequencing of genomes and cDNA libraries

bio-inorganic: Descriptive of biochemical materials that include elements, frequently metals, not popularly associated with them, e.g. Fe-S redox proteins, cobalamin, Zn fingers, the Fe-Mo centre of nitrogenase.

bioluminescence: The chemiluminescent emission of light by a living thing, e.g. firefly, certain fungi.

biomimetic: Descriptive of a compound or process that artificially simulates the function of a natural compound or process, e.g. a catalyst that employs the principles of enzyme specificity and rate acceleration.

biomimicry: The design of structural materials according to examples from biology, e.g. a polymer strengthened by carbon fibres, inspired by the insect exoskeleton, which is a protein matrix reinforced by chitin fibres.

biomolecular interaction analysis (BIA): The detection by a biosensor of a specific binding event for quantification of the ligand. One kind of BIA is *surface plasmon resonance*, the observation of polarization of light reflected from a surface coated with one of the binding partners. The change in polarization provides, in real time, a measure of binding of the second partner.

biophile: An element that has been enriched from its environment by organisms and which, therefore, appears as local deposits in the biosphere as a remnant of that bioconcentration

bioreactor: A vessel and ancillary equipment used for the growth of cells. The bioreactor may be designed to maintain temperature and levels of oxygen and nutrients and to monitor cell density, nutrient or metabolite levels

bioremediation: Processes that use the capabilities of micro-organisms to treat waste products that may be environmentally harmful and to render them innocuous. *see also* phytoremediation

biosensor: A device, especially an electrochemical device that detects, quantitatively and in real time, the presence of an anylate or some biological event (e.g. respiration, enzymic activity, binding to an antibody) and converts it into an electrical signal. (*see also* biomolecular interaction analysis (BIA); sniffer-patch)

biosphere: The zone of the earth in which life of some form is possible. It consists of the earth's surface, the oceans and atmosphere, and the upper several kilometers of the earth's crust

biosynthesis: The process by which a biological structure, especially a relatively simple structure, is formed by a sequence of enzymic reactions that starts from common metabolites or '*synthons*', e.g. the synthesis of haem from glycine and succinyl-CoA; the synthesis of steroids from acetyl-CoA. For the assembly of larger structures, such as membranes, via non-covalent interactions, the term *synkinesis* has been coined, with the subunits being termed *synkinons*.

biotrickling: A method for purification of polluted air by passing it over a bed of solid material on which pollutant-degrading organisms have been immobilized.

bi-product analogue (collected substrate inhibitor): An inhibitor of a hydrolase that incorporates structural features of both products of catalysis, and thus can bridge the S_1 and S_1' subsites. (*see also* specificity subsite)

bis-: A prefix derived from the Latin for twice.

bis-ligand: (*see* affinity precipitation)

bisnor-: A prefix that signifies a product of the removal of two-CH_2-groups.

bite-chew mechanism: (*see* proteasome)

biuret reaction: A colour reaction for the quantification of protein in solution. By analogy with the compound biuret (H_2N-CO-NH-CO-NH_2), the peptide backbone of proteins reacts with alkaline copper solutions to produce a violet colour.

bivalency: The property of an immunoglobulin G molecule, and some other immunoglobulins, of having two antigen-binding sites.

blast cell: (= stem cell (blast cell))

bleb: A small bud-like projection from the surface of a cell.

blood clotting cascade: The sequence of reactions, initiated by exposure of blood to extravascular surfaces, that results in a fibrin clot (*see also* extrinsic pathway; intrinsic pathway)

blood group substances: The oligosaccharide moieties of glycoproteins that appear in many biological fluids (saliva, urine, milk) as well as on the surface of erythrocytes. These antigens, upon reaction with specific antibodies, cause agglutination of the cells to which they are attached. Examples are the A, B and O antigens.

blotting: The transfer by contact of a macromolecule from a two-dimensional separation medium, e.g. paper or polyacrylamide slab, to another surface of higher affinity, e.g. nitrocellulose or nylon. (*see also* dot blotting; electroblotting; northern blotting; Southern blotting; south-western blotting; western blotting (immunoblotting))

blunt ended (flush ended): Descriptive of the structure of double-stranded DNA in which neither strand of the duplex extends further from the end than the other; often the product of cleavage by a restriction endonuclease. (*see also* sticky ended)

B-lymphocyte: (= B-cell (B-lymphocyte)).

boat form: (*see* chair form)

Bohr effect: The decrease in the affinity of haemoglobin for oxygen that occurs when the haemoglobin solution is made more acid above pH 6. The opposite occurs below pH 6; hence the physiological phenomenon is the *alkaline Bohr effect.*

boomerang DNA amplification (BDA): A technique similar to PCR that also uses a heat-resistant polymerase and cycles of polymerization, denaturation and annealing, but which requires only one primer. The source DNA is digested with a restriction endonuclease. A universal adaptor is engineered with self-complementary sections, so that it loops back upon itself and has ends that permit ligation to both strands at each end of the restriction fragment. In the first amplification cycle a primer anneals to an internal site and the polymerase copies the primer-binding strand through the adaptor, back along the second strand and past the site that is complementary to the primer-binding site of the first strand. In subsequent cycles the primer will have two sites for initiation of polymerization.

boron analogue: A derivative of an enzyme substrate in which; B-OH replaces; C=O of an amide or ester as an approximation to the transition state for its hydrolysis. (*see also* transition state inhibitor)

bottlebrush: (*see* proteoglycan (mucopolysaccharide))

boundary complex: (*see* insulator)

bouquet: (*see* non-collagen collagen)

bovine: Derived from cattle

bp: (= base pair)

BPS: Abbreviation for branch-point sequence.

brady-: A prefix derived from the Greek for slow.

branch migration: (*see* Holliday model)

branch site: In RNA splicing, the site to which the 5'-guanylate at the end of an intron is joined to an adenylate residue within the intron through a 2',5'-phosphodiester bond to form a 'lariat' intermediate. (*see also* self-splicing)

branch-point: A single metabolite that is an intermediate in two or more biosynthetic pathways, e.g. pyruvate (a precursor of acetyl-CoA, alanine and oxaloacetate), chorismic acid (a precursor of phenylalanine, tryptophan and tyrosine).

branch-point sequence (BPS): The sequence near the 3'-end of an intron of nuclear mRNA that contains the adenosine residue which, as the intron is excised, will accept the guanosine residue at the 5'-end of the intron. (*see also* pre-mRNA splicing)

breakthrough organism: A progenote; in early evolution, the first organism able to transmit genetic information, even if with only marginal fidelity.

bricolage: In evolutionary theory, the hit-or-miss recruitment and adaptation of pre-existing genes or materials for new functions

British anti-lewisite (BAL): 2,3-Dimercaptopropanol, an antidote for poisoning by arsenite which reacts with reduced lipoic acid.

brown adipose tissue: Thermogenic fatty tissue with a high content of relatively uncoupled mitochondria; especially prominent in infants, located around the kidneys and neck.

browning reaction (Maillard reaction): Covalent reactions that occur on heating proteins; the formation of cross-links between side chains and with carbohydrates.

brush border: The apical surface of intestinal epithelia, characterized by microvilli that extend into the lumen.

b-type ion: (*see* a-type ion)

bubble duplex: A non-hybridized region of an otherwise double-stranded DNA, named according to its appearance in electron micrographs

bubble: A widened segment of duplicating double-stranded DNA, visualized by electron microscopy, that is formed by two replication forks that travel away from the common origin of replication. The enlargement, or bubble, is a region in which the strands of the original DNA are separated and paired with newly synthesized DNA.

budding: The process of formation of vesicles from a membrane, e.g. the formation of transfer vesicles from the endoplasmic reticulum; also the asexual propagation of a yeast cell by mitotic division of the nucleus and pinching off of a daughter nucleus and some of the cytoplasm.

buffer: A proton-binding compound that, in the region of its pK_a, competes with water for added protons and makes the pH relatively insensitive to added acid.

buffer capacity: A measure of the ability of a solution to maintain its pH in the face of the addition of acid or alkali; a capacity of 1 when 1 mol of acid (or alkali) is added to 1 litre causes a pH fall (or rise) of 1 pH unit.

buffer value: (= buffer capacity)

buoyancy factor: In analytical ultracentrifugation, one of the factors that determine a macromolecule's rate of sedimentation; (1-*ur*), where *u* is the partial specific volume of the macromolecule (expressed as ml/g) and r is the density of the solution (expressed as g/ml).

bZIP protein: A type of Y-shaped homodimeric DNA-binding protein in which the stem is composed of two intertwined α-helices held together as a leucine zipper; the arms of the Y are the basic regions that reach around the double-stranded DNA and interact with its acidic phosphate moieties in a scissor-like grip. (*see also* helix-turn-helix (HTH) motif)

C

-clast: A suffix that indicates a cell that destroys or resorbs, e.g. osteoclast.

-cyte: A suffix with the same derivation as cyto-.

c: A prefix that denotes cellular. (*see also* oncogene)

C. elegans: (*see* Caenorhabitis elegans)

C kinase: (= protein kinase C (PKC))

C value: The calculated value of the total DNA content of a cell per haploid number of chromosomes.

C_3 photosynthesis: Carbon dioxide fixation by the reductive pentose phosphate pathway, i.e. the Calvin cycle. (*see also* C_3-C_4 photosynthesis; C_4 photosynthesis (C_4 cycle; Hatch-Slack pathway))

C_3-C_4 photosynthesis: A variant of C_3 photosynthesis in which the 2-carbon product of photorespiration is efficiently oxidized to CO_2, which is then recycled in photosynthesis. (*see also* C4 photosynthesis (C4 cycle; Hatch-Slack pathway))

C_4 cycle: (= C4 photosynthesis (C4 cycle; Hatch-Slack pathway))

C_4 photosynthesis (C_4 cycle; Hatch-Slack pathway): A variant of C_3 photosynthesis in which CO_2 is first concentrated at the site of photosynthesis by carboxylation of phosphoenolpyruvate to oxaloacetate, which is then transported as such or as an interchangeable 4-carbon dicarboxylic acid to the site of photosynthesis. This is an advantage to the plant because the key enzyme, ribulose bisphosphate carboxylase, which in photosynthesis forms 2 mol of 3-phosphoglycerate, has a competing activity, O_2 photorespiration, which produces both 3-phosphoglycerate and phosphoglycolate. (*see also* C3-C4 photosynthesis)

C_6C3 metabolite: (= shikimic acid metabolite (C6C3 metabolite))

Ca^{2+}-myristoyl switch : The mechanism by which Ca^{2+} signals a change in a myristoylated protein. In its non-liganded state, the acylated N-terminal amino group is buried in the protein's interior; upon binding Ca2+, the protein undergoes a dramatic conformational change, which exposes the acyl group and allows it to insert into a lipid bilayer membrane.

Ca^{2+}-myristoyl switch: The mechanism by which Ca^{2+} signals a change in a myristoylated protein. In its non-liganded state, the acylated N-terminal amino group is buried in the protein's interior; upon binding to Ca^{2+}, the protein undergoes a dramatic conformational change, which exposes the acyl group and allows it to insert into a lipid bilayer membrane.

CAAT box: A regulatory sequence upstream from some eukaryotic structural genes.

CAAX box: (*see* prenyl)

Caenorhabitis elegans: A nematode; a small worm favored by developmental and molecular biologists because if its ability to grow under laboratory conditions, its short generation time and its transparency. Because of its simplicity, it is possible to trace the development of the zygote to each of the approx. 1000 cells of the adult

cafeteria feeding: In laboratory animal nutrition, feeding a diet of low nutritional value, i.e. high in calories but low in protein and vitamins.

calcium-induced calcium release (CICR): A mechanism of cellular metabolic control: inositol 1,4,5-trisphosphate, generated in response to cell stimulation, causes a flow of Ca^{2+} across the plasma membrane, which in turn leads to periodic release of Ca^{2+} from the endoplasmic reticulum by an inositol trisphosphate-independent mechanism.

calorimetry: (*see* isothermal titration calorimetry)

calpain: A calcium-dependent cellular proteinase with a neutral pH optimum.

calpastatin: A protein inhibitor of a calpain.

Calvin cycle (reductive pentose cycle): The series of metabolic reactions by which carbon dioxide is fixed into glycolytic intermediates; the dark reactions of photosynthesis. (*see also* Hill reaction)

candidate: A compound, metabolite or peptide, etc. with a putative role; e.g. a candidate hormone may be a natural peptide which has *in vitro* bioactivity, but for which its characteristics *in vivo* have not yet been demonstrated.

canine: Derived from the dog

canyon hypothesis: A proposal to reconcile both the necessary invariability, and apparent non-immunogenicity, of the host cell receptor attachment site of a virus (e.g. a rhinovirus), and the ability of the same virus to mutate into many different immunologically recognized serotypes. The critical attachment site of the virus is sequestered at the bottom of a deep, narrow cleft, a 'canyon', which allows access to the receptor but not to an immunoglobulin; sites at the 'rim' of the canyon are accessible to immunoglobulins, but as they are not critical to the virus, they can mutate freely.

cap site: The site on a DNA template where transcription begins. It corresponds to the nucleotide at the 5'-end of the RNA transcript which accepts the 7-methylguanine cap.

cap: The7-methylguanine nucleoside attached to the 5'-end of mRNA by a 5'-5'-triphosphate bond.

capacitative Ca^{2+} entry: (*see* store-operated calcium entry)

capillary array electrophoresis (CAE): A method for rapid separation of large numbers of DNA fragments and their characterization by length, e.g. for identification of genetic markers for forensic or genetic linkage analysis. The mixture of fragments is labelled with a fluorophore for sensitive detection and separated under denaturing conditions by capillary electrophoresis. The sizes of the fragments are determined with the aid of standards of known length, which are labeled with different fluorophore

capillary electrophoresis: (*see* electrophoresis)

capping: The process of modifying the 5'-end of eukaryotic mRNA with 7-methylguanine. (*see also* cap)

caprine: Derived from the goat.

capsid: The protein coat of a virus.

captive antibody: (*see* sandwich immunoassay)

carbanion: A transient species in which a carbon atom bears a formal negative charge, e.g. an intermediate in the carboxylation of glutamate residues of preprothrombin. (*see also* carbocation (carbonium ion))

carbocation (carbonium ion): Any stable or transitory ion in which a carbon atom bears a formal positive charge. (*see also* carbanion)

carbohydrate: One of a class of biological materials comprising sugars, polymers of sugars, and compounds related to them. The name derives from the basic sugar structure, $(CH_2O)_n$. The category includes reduction and oxidation products, phosphate and sulphate esters, and amine derivatives.

carbon cycle: The movement of carbon atoms through different chemical forms and locations, from dissolved CO_2 in equilibrium with atmospheric CO_2 through plant carbohydrate, fats and proteins of plants and animals, and via oxidation back to atmospheric CO_2.

carbonium ion: (= carbocation (carbonium ion))

carboxy-: A prefix which refers to -COOH, the carboxy group.

carboxyl proteinase: (= aspartate proteinase)

carboxyl-terminal analysis: (= C-terminal (carboxyl-terminal) analysis)

Carboxy-terminal domain (CTD): The domain of a protein which includes the carboxy-terminal amino residue.

carcinoma: A malignant tumour of mesodermal origin.

cardiac glycoside: (*see* saponin)

cardiac muscle: A form of striated muscle characteristic of the heart.

cardio-: A prefix derived from the Greek for heart.

cargo: A sub-cellular structure that is transported by an intracellular trafficking system, especially secretory or membrane vesicles along microtubules or chromosomes along the mitotic spindle.

carotenoid: A derivative of phytoene, a symmetrical non-cyclic 40-carbon terpene.

carrier: Non-isotopically labelled material that dilutes a labelled tracer and enlarges the pool of the labelled metabolite, thus assisting in its isolation.

cartwheel: (*see* triplet)

cascade: A series of enzymic reactions that at each step convert an inactive enzyme into an active enzyme, which in turn activates another inactive enzyme, and thus greatly amplifies the initial signal. In a *homocascade* all the events involve one enzymic type (e.g. proteolysis); in a *heterocascade* the events involve different enzymic types (e.g. phospholipase and kinase).

caspase: Cysteinyl aspartate-specific proteinase; a member of a group of cellular cysteine endopeptidases that cleave so as to leave an N-terminal aspartate residue. The group includes interleukin-converting enzymes and enzymes associated with activation of apoptosis either as initiators near the head of a cascade of events, or as executioners, which perform limited proteolyisis of cellular proteins at the bottom of the cascade.

cassette mutagenesis: A method to determine the limits of tolerance of a protein to amino acid substitution. A cloned gene is mutated by synthesis of short segments of the gene with random base substitutions, insertion of these altered polynucleotide sequences into the gene and then transformation of cells with the mutated gene. Cells that contain a functional protein are selected and the protein (or gene) is examined to identify the successful mutation. Testing of many such short segments until the entire structural gene is examined maps the areas of tolerance and of sensitivity of the gene product to substitutions.

cassette: A DNA fragment that contains a non-coding sequence (for gene knock-out) or a mutated replacement of a normal gene, and flanked by sequences homologous to those flanking the normal gene. By homologous recombination of its flanking sequences with the target genome, the cassette will replace the normal gene, albeit at a low frequency. The cassette may also contain genes that allow detection, e.g. the LacZ operon, and selection, e.g. neosporin resistance.

CASTing: (= cyclic amplification and selection of targets (CASTing))

CAT assay: (= chloramphenicol acetyltransferase (CAT) assay)

cata-: A prefix derived from the Greek for down which refers to degradative processes. *see* ana-

catabolism: The action of energy-yielding metabolic pathways that degrade macromolecules and complex compounds or small molecules into CO_2, H_2O, etc. (*see also* amphibolic pathway; anabolism)

catabolite: A degradation product derived from a more complex compound. (*see also* anabolite)

catabolite activation: (*see* glucose effect)

catabolite repression: (*see* glucose effect)

catalytic antibody (abzyme): An antibody with catalytic properties; often raised against a hapten that chemically resembles the transition state of the intended substrate so as to force a substrate into that transition state.

catalytic configuration: (*see* entropy effect)

catalytic DNA: (*see* ribozyme)

catalytic rate constant (k_{cat}): The first-order rate constant that describes the rate-limiting step in enzyme catalysis, usually the conversion of the enzyme-substrate complex into the enzyme-product complex; the maximal velocity divided by the enzyme concentration. (*see also* turnover number (catalytic-centre activity))

catalytic site: The region of an enzyme that interacts with the substrate to effect the enzymic reaction.

catalytic subunit: (*see* regulatory subunit)

catalytic triad: (*see* charge relay system)

catalytic-centre activity: (*see* turnover number (catalytic-centre activity))

catecholamine: One of a family of phenolic compounds chemically related to catechol (1,2-dihydroxybenzene), which is derived metabolically from tyrosine; the family comprises hormones and neurotransmitters, including adrenaline (epinephrine), noradrenaline, dopamine, etc.

catenane: (= concatenate (catenane))

CATH: The acronym for class, architecture, topology, homologous superfamily; a hierarchical system of classification of protein structures named for the highest of the hierarchies: class refers to the secondary structure content (i.e. α-helix, β-sheet, both, little secondary structure); architecture refers to the motifs (e.g. ribbon, sheet, barrel); topology describes the overall shape and connectivity of the structural features; and homologous superfamily refers to the groups of proteins that have been classed together on the basis of amino-acid sequence similarities. Any protein can be assigned first to a class, then within the class to an architecture, then to an topology, etc.

cathepsin: An endopeptidase, often with low pH optimum, associated with lysosomes in the cell.

cathode: A ion attracted to the anode of an electrolytic cell, which being positively charged, attracts anions, which are, therefore, negatively charged. Similarly, a cation, which is attracted to the negatively charged cathode, is positively charged.

cation exchanger: (*see* ion-exchange chromatography)

caveolae: Small invaginations of the plasma membrane, characterized by glypiated proteins among others, that provide binding sites and points of entry of small molecules into the cytoplasm via trancytosis or potocytosis. *Trancytosis* is the pinching off of the ligand-bearing vesicle into the cytoplasm. *Potocytosis* is the putative closure of the narrow neck of the invagination and consequent isolation of the vesicular space from the extracellular fluid; the high local concentration of the ligand allows diffusion of trapped molecules into the cytoplasm.

ccDNA: Closed circular DNA.

CD: (= circular dichroism (CD))

cDNA: Complementary DNA; DNA that is synthesized, by reverse transcriptase, from an mRNA template, and therefore has no introns. (*see also* genomic DNA)

cDNA library: A collection of cells, usually *E. coli*, transformed by DNA vectors each of which contains a different cDNA insert synthesized from a collection of mRNA species. (*see also* genomic library)

CDR: (= complementarity-determining region (CDR))

cecropin: One of a group of inducible anti-microbial oligopeptides found in the haemolymph of a moth that serve some of the same functions as the immune system of higher species. (*see also* magainin)

cell culture: An *in vitro* technique for the propagation of genetically homogeneous, dispersed animal or plant cells in a differentiated state, usually in complex growth media. (*see also* explant)

cell pole: One of the two foci of a cell during mitosis, defined by a centriole, from which half the mitotic spindle radiates towards the other pole.

cell wall: The rigid, mainly cellulose, structure that surrounds the plasma membrane of a plant cell.

centimorgan (cM): A genetic measure of the distance that separates markers in the same chromosome, especially in describing a map of the chromosome; the distance that permits a 1% frequency of crossing over (1 morgan is the distance that permits 100% crossing over); equivalent to about 1310^{6} bases.

central dogma: In biology, the proposition that the permanent repository of genetic information is DNA which can be replicated, and that the information is expressed unidirectionally by transcription into RNA and thence by translation into protein; later qualified in important ways, e.g. upon the discovery of reverse transcription.

central nervous system (CNS): A division of the nervous system of higher animals that consists of the brain and spinal cord. (*see also* peripheral nervous system (PNS))

centrifugal elutriation: A method for separation of isolated cells according to their characteristic sedimentation rates in a centrifuge rotor which is designed to allow flow-through of a fluid during operation; also known as countercurrent elutriation and elutriation centrifugation. The centrifugal force on the cells is opposed by the force of a fluid moving in the opposite direction. Cells are first sedimented in a density gradient, then displaced by a buffer of increasing density that flows into the bottom of the sample cell and out of the top, and carries with it cells of the same density.

centriole: One of two cylindrical structures that appear in animal cells. During mitosis the centrioles are the poles from which the microtubules of the mitotic spindle radiate; they represent specialized microtubule-organizing centres.

centromere: The constriction of a chromosome seen in metaphase at which pairs of sister chromatids are held together and which is the site of attachment of a chromatid to the mitotic spindle.

centrosome: One of the structures of a cell that during mitosis serves as a microtubule-organizing centre; it becomes the structure from which the mitotic spindle radiates and which defines a cell pole.

cerebro-: A prefix derived from the Latin for brain.

chair form: A conformation of a six-membered non-aromatic ring that places all opposing centres away from each other, as contrasted with the *boat form*, which places two of the opposing centres towards each other. A *twist conformation* is a slightly flattened and relaxed variation of the chair form.

chalone: An inhibitory hormone.

channel former: A structure in a membrane that, without itself moving, allows passive diffusion of ions across the membrane.

chaotropic agent: A solute that disrupts the structure of the bulk water phase and, in so doing, changes the solubility and stability properties of other solutes, such as proteins.

chaperone machine: A multi-component system that ensures the proper folding of nascent proteins or their intracellular transport following synthesis; components include chaperones (eukaryotes) or chaperonins (from prokaryotes, mitochondria and plasmids), which bind to unfolded proteins, peptidyl proline isomerases and heat-shock proteins, which can serve various ancillary functions. (*see also* RNA chaperone)

chaperone: (*see* molecular chaperone; RNA chaperone)

chaperonin: A chaperone machine of a prokaryote mitochondrion or plasmid; a class of molecular chaperones, a complex of proteins which functions to properly fold denatured or improperly folded globular proteins. Chaperonins are typified by GroEL of *E. coli*, cylinders composed of two rings, each of seven subunits with 45 Å diameter chambers at each end. These identical chambers are lined with hydrophobic residues. One chamber accepts an unfolded protein substrate, which is characterized by a high proportion of exposed hydrophobic residues. In the case of GroEL, and possibly other chaperonins, the chamber is capped with another subunit (GroES), seven ATPs are hydrolyzed to ADP and Pi, and a conformational change in GroEL hides the hydrophobic residues of the chamber lining and presents polar residues. This is presumably the force that drives a refolding of the substrate so as to sequester in its interior its own hydrophobic residues. It is apparently the binding of an unfolded protein by the second chamber that drives uncapping of the first chamber and release of the refolded first protein.

Chargaff's rule: An empirical finding that in DNA the frequency of A equals the frequency of T, and the frequency of G equals the frequency of C; later given a theoretical basis by the Watson-Crick double-helix model of DNA.

charge: In chemistry, the integral sum of positive and negative particles that make up a species, a cation if the charge is positive, an anion, if the charge is negative. In biochemistry, the loading of a binding site with its intended ligand, e.g. the appropriate amino acid bound to its tRNA; in cell biology, essentially as in biochemistry, but on a macromolecular level, e.g. the charging of a liposome with material for transport cross a cell membrane.

charge accumulator model: A conceptualization of the mechanism of the water-splitting enzyme of photosynthesis in which four electrons are sequentially removed from the manganese centre, enabling it to oxidize water to molecular oxygen.

charge ladder: A series of elution peaks seen upon capillary electrophoresis of modified versions of a protein that vary only in their electrostatic charge. (*see also* electrophoresis)

charge relay system: A tautomeric form of a protein in which a formal charge on one dissociating group is moved to another in the same protein. In serine proteinases, for example, an internal β- (sometimes a α gamma-) carboxylate is partially protonated by bridging via an imidazole to an active-site nucleophile, often a serine hydroxy or cysteine thiol group, leaving the nucleophile with a negative charge and making it more reactive in its attack upon a substrate molecule. This group of residues is termed the *catalytic triad.*

checkpoint: A point through which the cell cycle cannot progress when its DNA is damaged or its chromosomes are not correctly attached or aligned.

chelator: A compound that binds a ligand, especially a metal ion, by several functional groups whose combined effect results in a high-affinity interaction; e.g. ethylenediaminetetraacetate (EDTA).

chemical coupling hypothesis: Explanation for oxidative phosphorylation by analogy with substrate-level phosphorylation, i.e. when a phosphate derivative of an electron carrier is oxidized, it becomes a high-energy phosphate compound, energetically capable of transferring its phosphate to ADP. (*see also* chemiosmotic theory; conformational hypothesis)

chemical force microscopy: A variant of atomic force microscopy in which the sensing probe is coated with a substance that interacts with the test material or object.

chemical mismatch detection: (= hydroxylamine and osmium tetroxide (HOT) technique (chemical mismatch detection))

chemical shift: In nuclear magnetic resonance, the modulation of field strength necessary to achieve the resonance frequency characteristic of a particular atom in a particular chemical, i.e. bonding, environment.

chemiluminescence: The light emitted by an exergonic chemical reaction; contrasted with *photoluminescence* (fluorescence or phosphorescence), which is the light emitted as a fluorophore falls to a lower energy state after having been excited by irradiation.

chemiosmotic theory: A proposed mechanism of oxidative phosphorylation in the mitochondrion and chloroplast that requires (1) electron transport to be arranged across the mitochondrial or chloroplast membrane so that protons are vectorially transported to its outer surface, (2) ATP synthesis to be arranged in the membrane so that the proton gradient can be used to drive ATP synthesis, and (3) that the mitochondrial/chloroplast membrane is impermeable to protons and defines an osmotically isolated space. The electron transport chain is composed alternately of hydrogen atom carriers and electron carriers, so that transfer from the former to the latter permits liberation of protons and their vectorial transport across the mitochondrial or chloroplast membrane. Because proton transport is not accompanied by an equivalent transport of electrons across the membrane, it generates not only a chemical potential (ΔpH) but also an electron potential ($\Delta\mu$), and the two together are termed the proton motive force. (*see also* chemical coupling hypothesis; conformational hypothesis; electrochemical gradient; P:O ratio; proton motive force)

chemokine: A chomotactic cytokine.

chemotaxis: The movement of a cell along the concentration gradient of a chemical, the chemotactic agent, toward its source, in the case of the chemical being a chemoattractant, or away, in the case of a chemorepellant.

chemotroph: An organism that derives its energy from chemical reactions, usually by the oxidation of nutrients by molecular oxygen. (*see also* phototroph)

chemotype: A strain of bacteria defined by variations in a biosynthetic product, e.g. Gram-negative bacteria chemotypes defined by the chemistry of their lipopolysaccharides.

chemzyme: A chemical enzyme; a soluble small organic molecule, sometimes with the addition of a transition metal ion, that is capable of catalysing specific reactions.

chiasma: The point of attachment of two chromatids during meiosis that results in *crossing-over*, which is the exchange of a region of one chromosome, from an end to the chiasma, with the homologous region of a sister chromosome.

chimaeric antibody: (See humanized antibody)

chimaeric DNA: A hybrid molecule produced by combining DNA from two different species into a single polynucleotide.

chimaeric protein: A protein product of a chimaeric DNA gene.

chimaeric toxin: (See immunotoxin (chimaeric toxin))

chip (oligonucleotide array): In biotechnology, a component of a device for screening genomic or cDNA for mutations, polymorphisms or gene expression. A chip is a small (a few centimeters on each side) standardized glass or other solid surface on which thousands of immobilized oligodeoxynucleotide probes have been synthesized or robotically deposited in a predetermined array, so that automated recording of fluorescence from each of the spots may score successful hybridizations. A chip may be designed for the detection of all known genes of a species (human, mouse, yeast), or selected specific sequences. (*see also* small molecule library; differential hybridization)

chirality: 'Handedness'; the formal orientation in space of stereoisomers or potentially stereoisomeric compounds. (*see also* CIP classification; meso-carbon; Ogston hypothesis; prochirality; R; symmetry)

chloramphenicol acetyltransferase (CAT) assay: A procedure for evaluation of the regulatory properties of eukaryotic promoter sequences. The CAT gene (which encodes an enzyme found only in bacteria) is used as a 'reporter gene' in that it is fused to a promoter sequence and introduced into a eukaryotic cell, where the ability of the promoter to cause the expression of the CAT gene is monitored by assay of the enzyme's activity; the assay may involve thin-layer chromatographic analysis of the conversion of [^{14}C]chloramphenicol into acetyl [^{14}C]chloramphenicol.

chloride shift: The movement of chloride ions into erythrocytes as bicarbonate exits, due to the generation by carbonic anhydrase of bicarbonate from carbonic acid inside the cells as they pass through peripheral tissues and take up carbon dioxide.

chloroplast: An organelle of a green plant cell in which light harvesting and ATP synthesis occur. (*see also* thylakoid membrane)

chole-: A prefix derived from the Greek for bile which refers to the gall bladder.

cholinergic: Responsive to acetylcholine.

chondro-: A prefix indicating cartilage, e.g. chondrocyte.

CHOnomics: (*see* omic research)

Chou-Fasman analysis: A method for prediction of the secondary structure of a protein from its primary structure. A survey of proteins of known secondary and tertiary structure allows an empirical classification of amino acid residues according to their abilities to strengthen or destabilize α-helix or β-structure. Recent advances have high-speed computers using Monte Carlo calculations to first fit multi-helix proteins into individual -helices, then use empirically assigned interaction coefficients to locate the helices in spatial relation to each other.

chromaffin granule: A dichromate-staining vesicle in cells that contains catecholamines or 5-hydroxytryptamine (serotonin); found for example in chromaffin cells of the adrenal medulla.

chromatid: One of a pair of chromosomes.

chromatin: The complex of DNA and associated proteins, most notably histones, that occurs in the nuclei of eukaryotic cells.

chromatogram: A graphical representation of a chromatographic separation, e.g. absorbance or radioactivity of the eluate (ordinate) plotted as a function of eluate volume (abscissa).

chromatophore: An epithelial cell of a lower animal in which pigment granules can be physically moved to effect colour changes.

chromatophoresis: A technique for protein separation that uses high-pressure liquid chromatography followed by sodium dodecyl sulphate (SDS)/polyacrylamide-gel electrophoresis. The effluent from a reverse-phase column is mixed with SDS and a reducing agent and applied to a polyacrylamide slab gel; the resulting gel shows two-dimensional separation, by polarity in one dimension and by molecular mass in the other.

chromo-: A prefix with the same derivation as -chrome.

chromomere: (*see* AT queue)

chromophore: A chemical group in a compound or macromolecule responsible for the absorbance of visible or ultraviolet light.

chromosomal banding: (*see* AT queue)

chromosomal map: The localization of genetic traits to regions of chromosomes, by linkage studies (e.g. the association of haemophilia with the X chromosome) and by *in situ* hybridization to fluorescent probes.

chromosome landing: In positional cloning, an alternative to chromosome walking for finding a gene. Clones of genomic DNA are fragmented so as to include both the target gene and a closely linked marker and are screened to select ('land on') those clones that contain the target gene.

chromosome painting: (See coincidence painting (chromosome painting))

chromosome walking: A strategy for mapping and sequencing a chromosome. Large restriction fragments are generated and, by Southern blotting with an RNA or cDNA probe, a single starting point is identified. New probes are synthesized from sequences of the same fragment that are adjacent to the starting point, and are then used to identify different restriction fragments adjacent to that which bears the starting point. The procedure is used repetitively, working towards the ends of the chromosome and away from the starting point.

chromosome: One of the nuclear structures, composed largely of chromatin, into which eukaryotic genes are organized.

CICR: (= calcium-induced calcium release (CICR))

cilium: A short extension of the plasma membrane with an axoneme core that functions in cell mobility or movement of extracellular substances across the cell surface.

CIP classification: The system of nomenclature of asymmetrical compounds devised by Cahn, Ingold and Prelog. (*see also* chirality; R)

circadian rhythm: The cyclical pattern of daily increases and decreases in enzyme and hormone levels and other physiological functions.

Circe effect: (*see* entropy effect)

circular dichroism (CD): A technique for determination of the asymmetry of a molecule. Unlike in a conventional polarimeter, where light is restricted to oscillation in a plane, in CD the direction of oscillation turns clockwise around the direction of the beam (right-circularly polarized) or counter-clockwise (left-circularly polarized). CD is expressed as $(A_L\text{-}A_R)/lm$, where for any wavelength A_L and A_R are the absorbances of left- and right-circularly polarized light respectively, l is the pathlength through a solution, and m is the molarity of a chiral solute. A CD spectrum of a protein in the far-ultraviolet, where the peptide bond absorbs, is characteristic of its conformation, whether *á*-helix, *â*-sheet or random coil. (*see also* optical rotation; optical rotatory dispersion (ORD))

cis: In stereochemistry, descriptive of a substituent on the same side of a structure that prevents equalization of positions by restricted rotation, e.g. an olefinic bond, a ring or a peptide bond; contrasted with *trans*, on the same side of the structure. In genetics, *cis* refers to linked markers and *trans* to separated markers.

cis-acting: Descriptive of the controlling effect of a regulatory gene on a structural gene that is adjacent to it; later broadened to distinguish intramolecular (*cis*) from intermolecular (*trans*) actions. (*see also* trans-acting)

cis-fusion: (*see* alpha (á)-configuration)

cis-Golgi: (*see* Golgi apparatus)

cis-recognition: (*see* trans-recognition)

cistron: A segment of DNA that contains all the information necessary for the production of a single polypeptide and includes both the structural (coding) sequences and regulatory sequences (transcription start and stop signals). (*see also* monocistronic mRNA; operon; polycistronic mRNA)

citric acid cycle: (= tricarboxylic acid cycle)

citrovorum factor: N^5-Formyltetrahydrofolic acid; originally isolated as a growth factor for *Leuconostoc citrovorum.*

clade: In phylogenetic taxonomy, families or subfamilies descended from a common ancestor.

clamping protein: (*see* SNARE hypothesis)

class I: (*see* major histocompatibility complex)

class II: (*see* major histocompatibility complex)

class II receptor: An acceptor; a membrane constituent that aids in transmembrane transport of nutrients, ions, etc.; distinguished from pharmacological receptors, which perform signal transduction, e.g. signal recognition followed by generation of a second messenger.

clathrate: A cage-like structure, e.g. that formed by water molecules that surround a hydrocarbon in solution.

cleft: The space between domains of a protein, often the binding or catalytic site of an enzyme. (*see also* pocket)

clonal selection theory: (See selective theory (clonal selection theory))

clone: In microbiology, a colony of cells all descended from a single ancestral cell and therefore genetically homogeneous; in molecular biology, an exact replica of a DNA fragment; *cloning* is the act of preparation and propagation of a genetically defined cell or a DNA fragment.

cloning of ligand targets (COLT): A strategy for screening phage cDNA libraries for genes that encode proteins that can bind a desired ligand. Phage which express the protein are identified on plates the ability to bind the labeled ligand.

clonomics: (*see* omic research)

cloverleaf: The formalized pattern assumed by a tRNA molecule viewed in two dimensions that shows the regions of internal complementarity that allow the polynucleotide to fold back upon itself into base-paired double helices. The stem includes the *acceptor stem* at the 3'-end, which attaches the amino acid, and the non-complementary loops or arms include the *anticodon arm*, which hybridizes with the codon of an mRNA. In three dimensions the structure can be divided into two sections at a right angle to one another: a coaxial stack that includes the acceptor stem, and the remainder of the molecule, i.e. the common arm that is shared by the two sections.

cluster analysis: In genomics, the discovery of genes that are similarly regulated. Statistical analyses are applied to the DNA microarray experiments to identify genes, the expressions of which respond in the same way to a variable, e.g. nutrient or hormone.

cluster of orthologous groups (COG): A collection of homologous genes that are useful for study of evolutionary relationships. A COG consists of orthologues (homologous genes that have diverged in different species from a common ancestral gene, along with the divergence of the species) and paralogues (genes in a single species that have arisen by duplication and divergence)

CMC: (= critical micelle concentration (CMC))

CNS: (= central nervous system (CNS))

coat protein: One of the proteins that encapsulate the nucleic acid core of a virus. (*see also* capsid)

coated pit: (*see* coated region)

coated region: The region of the plasma membrane of a cell capable of endocytosis that contains receptors and is lined on the cytoplasmic side with the protein clathrin. When the receptors are occupied, these regions invaginate to form *coated pits* and then pinch off into *coated vesicles* in the cytoplasm. (*see also* receptosome (endosome))

coated vesicle: (*see* coated region)

coatomer: A complex of four proteins that coats Golgi-derived non-clathrin-coated vesicles, and that presumably supports their function in intracellular transport and budding.

cochleate cylinder: (*see* vesosome)

cocktail: A premixed reagent, i.e. a solution of phosphors for scintillation counting.

code blocker (antisense drug): In an attempt to develop specific therapies, free of side-effects, an antisense oligonucleotide or analogue that is designed to form a double-stranded complex with an mRNA produced by a virus or other pathogen, or during inflammation or oncogenesis. In the absence of the blocker, the targeted mRNA would otherwise direct synthesis of a protein that supports the infection, inflammation or oncogenesis. (*see also* antisense)

coding joint: (*see* V(D)J recombination)

coding strand: (*see* antisense)

codon: The three-nucleotide sequence of an mRNA molecule that codes for one specific amino acid.

coenzyme I: Obsolete name for NAD^+.

coenzyme II: Obsolete name for $NADP^+$.

coenzyme: A co-substrate in some enzymic reactions that is usually present in limited quantities *in vivo* and which requires regeneration in subsequent reactions, e.g. coenzyme A, NAD^+, FAD.

cognate: A term borrowed from linguistics, signifying a correspondence; e.g. a receptor and its cognate ligand, a tRNA and its cognate amino acid.

cognisable: Descriptive of a ligand that is specific for a binding site, e.g. an inhibitor of an enzyme.

cohesive ended: (See sticky ended)

coiled coil: Several α-helices twisted together into a stout rope, e.g. in myosin, fibrin. (*see also* superhelix)

coincidence cloning: (*see* difference cloning)

coincidence painting (chromosome painting): A method for production and utilization of fluorophore-labelled oligonucleotides to identify very localized regions of chromosomes. The method of *degenerate oligonucleotide primed PCR* (DOP-PCR) generates an assortment of characteristic fragments by use of a primer with arbitrary but fixed 3'- and 5'-ends, but randomized internally, so that it will find an appropriate number of hybridization sites on both strands of a source DNA to generate a convenient number of PCR fragments; in a selection step, the DOP-PCR products are annealed to carefully selected and characterized chromosome fragments fixed to a solid support, washed to remove non-hybridizing fragments, and the successfully annealed oligonucleotides eluted and labelled to serve as 'paint' that will colour and thus identify the chromosome region that selected them.

co-linearity: The correspondence observed between the nucleotide sequence of a structural gene and the amino acid sequence of its protein product.

coliphage: (*see* bacteriophage)

collected substrate inhibitor: (= bi-product analogue (collected substrate inhibitor))

colligative: Descriptive of a property of a solution that depends upon the number but not the nature of solute molecules, e.g. osmotic pressure.

collisional limit: (*see* spare receptors)

colloid: A dispersion of a high-molecular-mass solute in a liquid phase. Typically, the colloidal solute is unable to traverse a semi-permeable membrane. The solute particles are usually of sufficient size to have an interior and a surface.

colony: The progeny of a single micro-organism grown in culture on solid media and visible as a spot on a plate.

colony hybridization: A technique for screening bacterial colonies for those that contain a desired polynucleotide sequence. A plate is exposed to a labelled oligonucleotide probe with a sequence complementary to part of the desired sequence which thus labels clones with that sequence. In a variant technique, *plaque hybridization*, phage vectors are screened for the desired polynucleotide sequence.

colostrum: The secretion of the mammary gland immediately following parturition, that differs from milk in its content of immunoglobulins and other specialized proteins.

colour complementation assay: (*see* colour PCR)

colour PCR: A technique for screening a population for a specific point mutation that uses fluorophore-labelled PCR primers. Three primers are designed for amplification of a restriction fragment that contains the mutation site near one end: the primer for the end distal to the mutation site is unlabelled; two primers for the other end containing the normal and the mutant sequence are each labelled with a different and distinguishable fluorophore, e.g. the normal with a red-fluorescing group and the mutant with a green-fluorescing group. When the restriction fragment is amplified in the presence of the primers, and the product separated from excess unincorporated primers, the amplified products will fluoresce red if from a normal subject, green if from a homozygous affected individual, and yellow if from a heterozygous individual.

combinatorial library: (*see* repertoire cloning; synthetic peptide combinatorial library (SPCL))

combinatorial: Descriptive of procedures or processes that generate a diversity of molecules (e.g. nucleic acid or protein sequences; related chemical structures) that can be screened for a desired property. The collection of molecules is termed a *library*. (*see also* small molecule library)

comet assay: Single cell gel electrophoresis; a method for assessing DNA damage in individual cells caused, for example, by apoptosis, radiation or chemical mutation. Cells embedded in an agarose gel are subjected to electrophoresis and stained with a DNA-binding dye. If the DNA is fragmented by double-strand breaks, small fragments will migrate toward the anode and under the microscope will appear as a central fluorescent body, the nucleus, and a smaller body, the fragments, giving the appearance of a comet head and tail. Alkaline treatment of the cells before electrophoresis will increase the sensitivity of the assay by denaturing the DNA, thus allowing detection of single-strand breaks.

co-metabolism: In microbial physiology, the metabolism of a xenobiotic which does not support growth. The phenomenon possibly involves the fortuitous metabolism of material due to partial non-specificity of key enzymes.

committed step: The first irreversible enzymic reaction in a metabolic pathway, usually controlled by an allosteric enzyme.

comparative anchor tagged site (CATS): An extended DNA sequence that is sufficiently conserved across a spectrum of species, e.g. all mammals or mouse and human, so as to serve as a marker for comparative genome mapping and assist the search in the genome of a second species of a gene for a trait known from a first species' genome. The most useful CATS will include coding sequences to assist their identification. (*see* expressed sequence tag; sequence tagged site)

comparative map: An application of the comparative mapping of genes and other genetic markers in the genomes of several mammalian species to the localization to a specific chromosomal locus of a gene, in particular one thought to be related to a genetic disease (a candidate disease gene). Because mammalian evolution has resulted in relatively few gene relocations, and since those that have occurred seem to be random, knowledge of linkages of genes in mammalian species may be helpful in locating a gene in another mammal's genome.

compartment: A functionally isolated reservoir of a metabolite; the space occupied by a pool.

competitive binding assay: A technique for assaying a substance by observing how effective it is, compared with standards, in displacing a fixed amount of the labelled (often radiolabelled) material from a binding site of great specificity and high affinity, usually an antibody or cell surface receptor. (*see also* ELISA; radioimmunoassay (RIA))

competitive inhibition: A form of enzyme inhibition in which the inhibitor competes with the substrate for the enzyme's substrate-binding site. The result is an increase in the K_m value while leaving V_{max} unaltered. (*see also* inhibitor; non-competitive inhibition (mixed inhibition); uncompetitive inhibition)

competitive labelling (differential chemical modification): A method for characterization of the microenvironment of an amino acid residue of a protein, e.g. its accessibility to solvent or its pK_a; other residues of the same amino acid serve as internal controls. Rates of reaction or ratios of reaction compared with control residues are observed; very limited amounts of reagent and extents of reaction maximize the competition for the reagent between similar groups on the protein and ensure that only unmodified protein reacts with reagent.

competitive oligonucleotide primary-utilized primers: COP-utilized primers; a method essentially the same as PCR amplification of multiple specific alleles, except the mismatches in primers occur in the middle of the primers.

competitive PCR: A variation of PCR to make it a semi-quantitative assay for a polynucleotide. A new restriction site is introduced into the polynucleotide to be assayed so that, when an unknown quantity of the original and a known amount of the modified template are used together, both will be copied but their products will be distinguishable. The products are quantified on the basis of binding a radiolabelled hybridization probe and incorporation of labelled primer or fluorescence of bound ethidium bromide; the ratio of their products is equated with the ratio of the original templates.

complement system: A series of blood proteins that contribute to innate immunity. When triggered by a foreign substance or particle, a cascade of enzymatic activations and reactions results in chemotaxis, activation of the acquired immune system and opsonization of the substance or particle to allow its clearance by phagocytosis and lysis. *see* macrophage; opsonin

complementarity-determining region (CDR): A polypeptide sequence of a variable domain of an immunoglobulin that is particularly responsible for its recognition by lymphocytes. These short sequences interrupt and loop out from the *framework regions* (FRs), which are relatively invariant and form the basic structural β-sheet scaffolding of the domains.

complementary base: The purine that can form hydrogen bonds with a pyrimidine, and vice versa, in a double-stranded polynucleotide, e.g. G with C and A with T. (*see also* complementary)

complementary: In nucleic acid chemistry, descriptive of the relationship between two polynucleotides that can combine in an antiparallel double helix; the bases of each polynucleotide are in a hydrogen-bonded inter-strand pair with a complementary base, A to T (or U) and C to G. In protein chemistry, the matching of shape and/or charge of a protein to a ligand.

complementation cloning: A method for identification of a DNA fragment and selection of the defective cell line upon which it confers resistance to a selective pressure, e.g. cells that are hypersensitive to ionizing radiation. Fractions of normal genomic cDNA are screened for ability to confer resistance on cells into which they are introduced. Different lines of defective cells that represent discrete genetic defects may be associated with different cDNA fragments.

complex carbohydrate: An oligomer or higher polymer of more than one kind of sugar moiety, or a glycoside formed with a non-sugar compound. (*see also* glycolipid; glycoprotein; glycoside; heteropolysaccharide; proteoglycan (mucopolysaccharide))

complexity: In nucleic acid chemistry, a measure of the unique sequences in a DNA preparation; it increases with chain length and decreases with the extent to which the sequences are repetitive; evaluated as the size of a DNA fragment with no repetitive sequences that has the same $C_0t_{1/2}$ value. In protein chemistry it is a quantitative measure of the randomness of a polypeptide sequence, *C*, which ranges from 0 for a homopolymer to 1.0 for a perfectly random sequence. For example, the complexity of collagen is 0.23 and that of globular proteins is about 0.90; $C=f_i \times \ln(f_i)$, where f_i is the mol fraction of amino acid *i* among all the amino acids represented in the protein. (*see also* Cot unit; simplified)

complex-type carbohydrate: One type of glycoprotein moiety that is attached to the â-amide nitrogen of an asparagine residue (N-linked) and which may contain sialic acid residues at the non-reducing ends. (*see also* hybrid-type carbohydrate; high-mannose-type carbohydrate)

compound heterozygous: (*see* allele)

concatenate (catenane): A structure formed by two or more interlinked closed circular (cc)DNAs; formation occurs between the rapid replication of a ccDNA and the slow nicking of one of the ccDNAs to release it from the others.

concatenation cDNA sequencing (CCS): (*see* double adaptor method)

concerted reaction: A chemical or enzymic reaction in which several operations, e.g. attachment of an incoming group and departure of a leaving group, occur simultaneously and on opposite sides of the reaction centre.

condensation reaction: The formation of a carbon-carbon, carbon-nitrogen or carbon-sulphur bond.

conditional expression: The technique of constructing cell lines, or transgenic animals, in which the expression of a gene may be controlled by the level of an exogenous compound. One such system, tet-on, involves a plasmid with the *Escherichia coli* tetracycline-dependent operator upstream from the coding sequence of the targeted gene: in the presence of a tetracycline, the gene is expressed. In the tet-off system, a regulator plasmid is also present, and encodes the tetracylcine-controlled transactivator protein, which brings the tetracycline operator into proximity with the promoter sequence of the gene and allows gene expression to proceed. If, however, a tetracylcine is present, expression is interrupted.

configuration: An arrangement in space at an asymmetrical centre. (*see also* alpha (á)-configuration; conformation)

conformation: For a compound or macromolecule that has at least limited freedom of rotation about its chemical bonds, one alternative arrangement in space of its constituent atoms and groups. (*see also* configuration; tertiary structure)

conformational change: The adjustment of a protein's tertiary structure in response to external factors (e.g. pH, temperature, solute concentration) or to binding of a ligand.

conformational hypothesis: An explanation for oxidative phosphorylation that proposes that electron transport builds up a store of energy in the distorted shape of the mitochondrial inner membrane, the relaxation of which is coupled with phosphorylation of ADP. (*see also* chemical coupling hypothesis; chemiosmotic theory)

congenic: Descriptive of strains of the same species, the genomes of which, vary only in a small chromosomal segment.

conjugation: In organic chemistry, the interaction of double or triple bonds or aromatic groups, separated by a single bond, which delocalizes their non-bonding electrons; e.g. -CH=CH-CH=CH-, -CH=CH-CH=O. In molecular biology, conjugation is the transfer of DNA between cells, usually bacteria, by cell-to-cell contact. (*see also* transduction; transformation)

connective tissue: Support tissue that contains extracellular fibres such as collagen to give strength and protection to blood vessels, nerves, etc.; also bone and cartilage.

connexon: (*see* hemi-channel (connexon))

Connolly surface: (*see* solvent-accessible surface)

consensus sequence: The minimal common sequence that appears in homologous polynucleotides or proteins, or limited regions of these, e.g. the sequence around the active-site region of serine proteinases.

conservation: In molecular biology, the preservation through time of some bases in the polynucleotide sequence of an evolving gene or of some amino acids in the sequence of an evolving protein.

conservative mutation: Such a mutation results in an amino acid substitution that preserves an essential chemical characteristic of the original, e.g. a leucine for an isoleucine, an aspartate for a glutamate, a lysine for an arginine.

conservative site-specific recombination (CSSR): The inversion or excision of a DNA sequence or the integration of a foreign DNA sequence into a homologous genomic site. These all require homologous recombination at the sites of inversion, excision or integration, and contrast with transposition, in which a DNA sequence is inserted or removed from a non-homologous site by excision of the donor sequence from its source, cutting open the acceptor site, and insertion of the transposed sequence and its ligation in place.

constant region: The part of the amino acid sequence of any one immunoglobulin that is the same for all immunoglobulins of its class. (*see also* hypervariable region; variable region)

constitutive enzyme: An enzyme that is produced at a constant, non-inducible, rate. (*see also* inducible enzyme)

contact inhibition: The cessation of division that occurs when cells in culture reach confluence and establish gap junctions with neighbouring cells.

contact number: A measure of the location of an amino acid residue in a protein molecule; the number of á-carbons, excluding those of the adjacent residues in the linear sequence, within an 8 Å radius of the á-carbon of any given residue; this varies from 0 for residues on a protein's surface to about 15 for those in the interior. Because there is a higher proportion of internal to external residues in larger globular proteins, the values can be normalized by calculation of a *relative contact number*, i.e. the difference between the average contact number for all residues of a protein and that of any given residue, which varies between -5 and +5.

contact site: A small region where the mitochondrial inner and outer membranes touch, and at which structures exist that are responsible for transport of proteins and adenosine nucleotides into and out of the mitochondrion.

contact system: The cascade of proteolytic activations of blood factors that is initiated by autoactivation of factor XII when it is adsorbed to a surface, e.g. that of a bacterium, and results in uncontrolled hypotension (shock) due to bradykinin release.

contig map: A characterization of a chromosome or large portion of one by *contiguous sequences*, which are zones of overlap of partially characterized segments; especially used as a strategy for co-ordination of efforts in many laboratories to sequence the human genome. (*see also* linkage map; physical map; restriction map; sequence-tagged site)

control: (*see* regulation)

convertase: A cellular proteinase that processes hormone precursors by recognition of the precursor sequence pairs of basic residues, which are the sites of cleavage.

converting enzyme: (*see* convertase)

co-operativity: (*see* negative co-operativity; positive co-operativity).

COP-utilized primers: (*see* competitive oligonucleotide primary-utilized primers) (*see also* PCR amplification of multiple specific alleles)

core glycosylation: The attachment to a protein and modification of N-linked carbohydrate moieties that occurs in the endoplasmic reticulum (*see also* terminal glycosylation)

core promoter: (*see* promoter)

core protein: One of the proteins to which are attached the carbohydrate chains (e.g. chondroitin sulphate, keratan sulphate) of a proteoglycan, and which with them comprise the subunits of the proteoglycan. (*see also* link protein)

Cori cycle: The transport of the precursor and product of glycolysis between exercising muscle and the liver, i.e. lactic acid from muscle to liver, and glucose from liver to muscle.

Cori ester: Obsolete name for a-glucose 1-phosphate.

Cornish-Bowden plot: A graphical method for determination of the type of enzyme inhibition and the dissociation constant for an enzyme-inhibitor complex (K_i) or for an enzyme-inhibitor-substrate complex (K_i'). The effect on the enzymic rate (v) is determined at two or more substrate concentrations (S) and over a range of inhibitor concentrations (I). In a plot of S/v versus I, the data for each substrate concentration fall on straight lines that intersect at I=-K_i and S/v=K_m/V_{max} (uncompetitive inhibition), or intersect on the abscissa (S/v=0) at I=-K_i (non-competitive inhibition). For competitive inhibition, the lines are parallel. (*see also* Dixon plot; Easson and Stedman plot; Hunter and Downs plot)

correlation spectroscopy (COSY): A two-dimensional nuclear magnetic resonance technique for detection of connectivity of the resonance of two centres through the covalent bonds that join them. (*see also* nuclear Overhauser and exchange spectroscopy).

correlation time: In electron spin resonance and nuclear magnetic resonance spectroscopy, a time constant for the interaction of an observed species with its environment, often related to the rate of decay of its observed property; affected by spin relaxation, rotational motion and chemical exchange.

corrin ring: A porphyrin-like ring system that lacks a methene bridge between two adjacent pyrrole rings; the basic structure of vitamin B_{12} and the coenzymes derived from it.

cortex: The outer layer of a structure, e.g. adrenal cortex, cell cortex. (*see also* medulla)

corticoid: An adrenal cortical steroid.

cosmid walking: A strategy for sequence analysis of a large polynucleotide; chromosome walking using a cosmid.

cosmid: A plasmid used to introduce DNA sequences that are much larger than is suitable for other vectors.

co-suppression: The inhibition of gene expression by both sense and antisense RNAs.

COSY: (= correlation spectroscopy (COSY))

co-translational: Descriptive of a process that occurs at the same time as protein synthesis; e.g. folding into a native conformation assisted by a molecular chaperone, transport of a secretory protein across the endoplasmic reticulum.

co-transport: The concerted transport of metabolites or ions across a membrane. (*see also* antiport; symport)

Cotton effect: A feature of some optical rotatory dispersion spectra that is usefully correlated with conformational and structural features of a molecule. A sharp peak in optical rotation followed by a deep trough is seen as the wavelength decreases; the wavelength between the peak and the trough where the rotation is zero is the point of maximal absorbance. A *negative Cotton effect* is the reverse, i.e. a fall in rotation followed by a steep rise as the wavelength decreases.

countercurrent chromatography: A variant of partition chromatography that does not use a solid support to stabilize the stationary phase. A horizontal helical tube is filled with the stationary phase which has been equilibrated with the mobile phase; the mobile phase with solutes to be separated is introduced into one end of the helical tube and fractions are collected from the other. Equilibrium between the phases may be enhanced by slow rotation of the tube around its axis.

countercurrent distribution: A method for separation of compounds based on differing partition ratios between immiscible liquids. In a linear series of many units (each consisting of an upper and a lower phase) after equilibration of the materials to be separated between two phases of one unit, the upper phase is repeatedly passed along for re-equilibration with the adjacent lower phase; this creates, after many such separations, binomial distributions of each constituent of the system among the upper and lower liquid phases of the units.

countercurrent elutriation: (= centrifugal elutriation)

coupled phosphorylation: The synthesis of the phosphate anhydride bonds of ATP using energy derived from the electron transport chain.

coupled plasmon-waveguide resonance (CPWR): (*see* surface plasmon resonance)

coupled-enzyme assay: The determination of a substrate or enzyme activity by coupling of one enzymic reaction with another, more easily detectable, reaction. The product of the first reaction is the substrate for the second. These assays are often designed to generate or consume a reduced nicotinamide-containing nucleotide, which is measurable by its ultraviolet absorption.

coupling between conformational fluctuations: The acceleration of a reaction that occurs when reactive groups are constrained in a productive orientation, either intermolecularly, as on an enzyme surface, or intramolecularly as in a model compound. Depending upon which aspect of this phenomenon is being emphasized, the physical bringing together of the components or some special feature of their association, it has been referred to variously as antichimaeric assistance, approximation, catalytic configuration, the Circe effect, coupling between conformational fluctuations, directed proton transfer, electric field effect, electrostatic stabilization, FARCE (freezing at the reactive centres of enzymes), gas-phase analogy, group transfer hydration, oribital steering, propinquity effect, proximity effect, rotamer distribution, stereoelectronic control, stereopopulation control, substrate anchoring, togetherness, torsional stress and virbrational activation..

coupling factor: A protein that permits the synthesis of ATP driven by the energy made available by mitochondrial electron transport.

covalent catalysis: A stage in some enzymic reactions in which one moiety of a substrate is attached by a covalent bond to the enzyme.

covalent modification: As applied to enzymes, the regulation of activity by modifications that may be reversible (e.g. phosphorylation or adenylation) or irreversible (e.g. limited proteolysis). (*see also* post-transcriptional modification; post-translational modification)

CpG island: A microsatellite sequence, a repeated CG sequence in genomic DNA, sometimes the sites of methylations.

Crabtree effect: The decrease in cellular respiration caused by increased glucose concentrations. (*see also* Pasteur effect)

crenarchaeote: (*see* eukaryote)

crenote: (*see* eukaryote)

Crick strand: A designation of one arbitrarily chosen strand of double-stranded DNA to distinguish it from the other, called the *Watson strand.*

crista: A deep indentation of the mitochondrial inner membrane.

critical micelle concentration (CMC): The minimum concentration of a detergent at which it will form micelles and below which it is a true solution.

crossing-over: (*see* chiasma)

cross-link: In protein chemistry, a natural or synthetic covalent bond between protein side chains; it can be directly between them or mediated by a spacer group. In carbohydrate and nucleic acid chemistry, a synthetic bridging group.

crossover: The physical exchange of homologous parts between a pair of individual chromatids.

crossover connection: A short polypeptide sequence that connects strands of b-structure.

crossover element: A DNA sequence that is a preferred site of crossover between two genes or two chromosomes.

crossover experiment: A method to detect the control point in a metabolic pathway by observation of the effect on metabolite levels as new steady-state levels are established after sudden creation or relieving of an experimental deficiency in a controlling factor, e.g. the effect of oxygen on the redox state of reduced electron transport components, or the effect of oxygen on the levels of glycolytic intermediates of anaerobic tissue.

cross-reacting material: A protein product resulting from mutation that has lost its function but is recognizable by its ability to react with antibodies raised against the normal protein. More broadly, material may cross-react because it bears an epitope in common with the antigen.

cross-regulation: Control of a metabolic pathway by the product of a different but related pathway, e.g. activation of the reduction of ADP to dADP by dGTP.

cross-screening: A method for reconstructing the order of random clones in a contig by identification of pairs with overlapping sequences. The DNA clones are deposited as parallel strips on a membrane, then strips of the corresponding RNA probe, which are labelled for detection, are laid over and perpendicular to the DNA strips. After hybridization and washing, hybridization is detected for identification of those pairs which share a sequence.

cross-species chromosome painting: (*see* zoo-fluorescence *in situ* hybridization)

cryptic: Descriptive of a function that is not usually, or not always available or expressed; e.g. an enzyme that is separated from poteintial substrates by being located in a lysosome, a splice site in a pre-mRNA that is used only under certain physiological conditions.

crypto-: A prefix derived from the Greek for hidden.

cryptochrome: A blue light- and ultraviolet-A-sensitive receptor protein, containing pterin and flavin prostehetic groups; found ubiquitously in plants and in mammals, and linked to the circadian response of these organisms to light.

cryptogene: (*see* guide RNA)

C-terminal (carboxyl-terminal) analysis: A chemical method for determination of C-terminal residues of proteins by cleavage by hydrazinolysis, which generates amino acid hydrazides of all amino acids except the C-terminal one; also an enzymic method using limited cleavage by a carboxypeptidase that sequentially liberates amino acid residues from the C-terminal position, i.e. the C-terminal residue first, the penultimate residue second, etc.

C-terminal: In a polypeptide sequence, that unique residue which is connected to the linear sequence by its amino group, leaving it with a free carboxy group. In practice, the carboxy group of a C-terminal residue may be modified, e.g. by amidation or, in the case of pyroglutamate, by internal lactamization. (*see also* N-terminal)

C-terminus: The carboxyl end of a protein. As proteins are linear polymers of -amino acids H_2N-CHR_a-CO-(NH-CHR'-CO)$_n$-NH-CHR_z-COOH where R is one of the 20 chemical groups that distinguish the amino acids that occur in proteins, the N-terminus is the amino acid residue H_2N-CHR_a-CO-, and the C-terminus is the residue -NH-CHR_z-COOH.

CTL: Cytotoxic T-cell. (*see* T-cell)

cubane: (*see* iron-sulphur protein)

culture: An inoculum of cells, especially a pure strain, intended for propagation in liquid or on solid media; also the act of propagation of the cells.

cut: Cleavage of the phosphodiester bonds of both strands of double-stranded DNA to produce blunt-ended fragments.

cycle sequencing: A method that recruits PCR technology to assist sequencing of small amounts of a single-stranded DNA template. A heat-stable polymerase, a single primer, deoxynucleotides and small amounts of one of the four labelled dideoxynucleotides are incubated with the template in each of four incubations; in each of a series of cycles of annealing, polymerization and denaturation, a spectrum of single-stranded DNA fragments is generated that can be electrophoretically separated to produce a ladder from which can be assigned nucleotide positions relative to the primer.

cyclic amplification and selection of targets (CASTing): Essentially identical to the *selected and amplified (protein) binding site oligonucleotide* (*SAAB*) and *target detection assay* (*TDA*) procedures; a procedure for identification of consensus sequences of DNA to which a protein, e.g. a transcription factor, may bind. A random polynucleotide sequence is synthesized flanked by two defined sequences that will serve as templates for PCR primers; the polynucleotides are exposed to the DNA-binding protein, any complex that is formed is separated from the unliganded polynucleotides (e.g. by gel shift assay, affinity chromatography, filter binding) and the polynucleotide of the complex is isolated and amplified by PCR; repeated recycling through the sequence of ligand formation, selection and amplification results in a preparation that is sufficiently pure to be cloned into bacteria for larger-scale production. A variant is *systematic evolution of ligands by exponential enrichment* (*SELEX*) for identification of RNA sequences, which begins with a mixture of polyribonucleotides and in each cycle produces DNA from the selected RNA-protein complex using reverse transcriptase, amplifies it by PCR, and then produces new RNA transcripts for the next round of selection.

cyclic nucleotide: An internal nucleoside phosphodiester; usually a 2',3'-diester or a 3',5'-diester.

cystatin: A term applied both to the superfamily of cysteine proteinase inhibitors and to one subgroup; the other subgroups are the kininogens and stefins.

cysteine proteinase (thiol proteinase): A type of peptidase that has at its active site a cysteine residue. (*see also* aspartate proteinase; metalloproteinase; serine proteinase)

cysteine scanning: The use of site-directed mutagenesis to determine, for residues of a complex, usually an uncrystallized protein, their proximity to other residues of the protein, their mileiu (extracellular, cytoplasmic, transmembrane) or the participation of the residues' side chains in specific interactions. Every residue of a protein is systematically replaced by a cysteine residue. Being of a moderate size so that physical disruption is unlikely to cause changes in function, and being easily converted into a useful reporter group by chemical modification, cysteine is a generally useful residue for modification.

cysteine switch: (*see* Velcro mechanism (cysteine switch))

cyto-: A prefix derived from the Greek for vessel which refers to the cell.

cytoflow: An instrument for flow cytometry.

cytokinesis: The separation of two daughter cells in the final stage of mitosis.

cytoplasm: The part of a cell inside the plasma membrane and outside the nucleus, comprising membrane-bound organelles, cytoskeleton, ribosomes and cytosol.

cytoplasmic streaming: Concerted internal movement in some large cells in which bands of cytoplasm move just under the cell surface.

cytoskeleton: The network of relatively rigid structures within a cell that give it shape and provide a framework against which intracellular movement may take place; includes microtubules, intermediate filaments, F-actin filaments and associated proteins.

cytosol: The soluble part of a cell's cytoplasm, i.e. that part that does not sediment during ultracentrifugation.

cytosolic receptor: As distinguished from cell surface receptors, an intracellular receptor protein, especially one for a steroid hormone. (*see also* nuclear receptor)

cytotoxic T-cell: (*see* T-cell).

D

2-D crystallography: A technique for low-resolution (10–20 Å) determination of shapes of proteins, especially large, complex proteins that are not amenable to crystallization by conventional techniques. A monomolecular layer of a lipid molecule is spread over an aqueous solution of the protein. A ligand that binds and orients the dissolved protein extends from monomolecular layer into the solution. The oriented two-dimensional protein molecular array is then analysed by electron diffraction crystallography.

D-: A symbol that indicates the absolute configuration of some asymmetrical metabolites, especially amino acids and sugars, and relates them to the configuration of (*d*)-glyceraldehyde. The opposite configuration is designated L-. α-Amino acids are related to glyceraldehyde as follows: hydrogen to hydrogen, carboxy to aldehyde, amino to hydroxy and side chain to methyl group. Sugars are classified according to the asymmetrical centre most remote from the carbonyl group of the molecule, e.g. C-5 of glucose. Note that absolute configurations are indicated as D- and L-. *d-* (*dextro*) and *l-* (*laevo*) refer to the direction of rotation of plane polarized monochromatic light, usually at 589 nm.

D-loop: A structure of heteroduplex polynucleotides in which one strand is longer than the other, forcing the internal extra nucleotides into a loop. (*see also* heteroduplex mapping; telomere)

dAB: (= single-domain antibody (dAB))

dalton (Da): The unit of measurement of molecular mass, in g/mol.

Danielli-Davson model: A model for biological membranes; a variation of the Gortner and Grendel model in which globular proteins are tightly adsorbed to the polar groups of the exposed surfaces of the membrane. (*see also* fluid mosaic model (Singer-Nicolson model); unit membrane Danielli-Davson model)

ddNTP: Dideoxynucleoside triphosphate. The letter N refers to any or all of the common bases.

de novo synthesis: The biosynthesis of a compound or chemical group from dissimilar compounds or groups, e.g. the formation of methyl groups from formyl groups.

DEAD box protein: One of a group of proteins that have in common an Asp-Glu-Ala-Asp (DEAD in the one-letter code) or related sequence, and function in ATP-dependent processing of RNA.

deamination: The abstraction of the elements of ammonia from a compound, e.g. from histidine by the histidine lyase reaction, or from AMP in the adenylate deaminase reaction.

death-associated protein (DAP): A protein that induces apoptosis. It binds to a death receptor on the plasma membrane, resulting in an intracellular cascade of events that acitivates caspases and eventually causes apoptosis.

death domain: A sequence of a death-associated protein or its gene, which is common to several death-associated proteins.

death gene: A gene whose expression is associated with apoptosis, e.g. a Ca^{2+}-activated endonuclease that cleaves exposed regions of chromatin to produce nucleosome-sized fragments. (*see also* oncogene)

death receptor: (*see* death-associated protein)

decorated thin filament: A thin filament of muscle, i.e. F-actin, to which are attached globular heads. The S1 fragments of myosin resemble in electron micrographs a string with attached arrowheads, all angled towards one end of the string.

decoy complex: *see* **insulator**

degeneracy: Redundancy of the genetic code, in that each amino acid is specified by more than one codon.

degenerate oligonucleotide primed PCR (DOP): (*see* coincidence painting (chromosome painting))

dehydro-: A prefix that signifies a product of the removal of two hydrogen atoms, e.g. dehydroepiandrosterone.

dehydrogenation: The oxidation of a compound by removal of equal numbers of protons and electrons, usually two of each.

delayed-early stage: (*see* immediate-early stage)

deletion map: The linear arrangement of natural mutations along a chromosome or part of a chromosome. In the first phase many individuals are studied for the presence of cytogenetic markers; later phases use restriction fragment length polymorphisms to detail the chemical nature of the aberrations.

deletion mapping: The assignment of the relative positions of deletion mutations along chromosomes. In the first phase, many individuals are studies for the presence of cytogenetic markers and later phases use restriction fragment length polymorphisms to detail the chemical nature of the aberrations. Futhermore, an experimental approach for identifying the location of a point mutation by testing whether recombination with a gene with a known deletion will produce the wild-type protein, in which case the site of point mutation lies outside the deleted region, or whether it fails to produce the wild-type protein, in which case the point mutation lies within the deleted region. A sufficiently large number of deletion mutants will narrow the location of the point mutation to a single nucleotide.

deletion mutation: A mutation caused by the absence of one of more nucleotides in the DNA sequence.

delta restriction cloning: A method for the use of a single primer to determine a DNA sequence longer than would normally be practical for a single sequencing attempt. After the first sequencing the bulk of the determined sequence is excised and the plasmid re-ligated, so that the primer is just upstream from the remaining undetermined sequence. The procedure may be repeated indefinitely until a large sequence is determined.

denaturation: The destruction of the ordered folding of a protein or nucleic acid that is required for its normal function. Protein denaturation often involves a change from a specific globular or fibrous conformation to a random coil; nucleic acid denaturation often involves the dissociation of a duplex into single strands. (*see also* native structure)

denaturing gel electrophoresis: (= sodium dodecyl sulphate/ polyacrylamide-gel electrophoresis (SDS/PAGE))

denaturing gradient gel electrophoresis (DGGE technique): A method for detection of single base substitutions in DNA fragments. Putative mutant and normal double-stranded DNA fragments are applied along one edge of an agarose gel slab that contains a denaturing agent (e.g. urea and formamide) in a gradient perpendicular to the direction of electrophoresis. At a low concentration of the denaturant the fragments are more mobile, and at high concentration they are unfolded and therefore less mobile. The electrophoretic band describes a sigmoid curve on the developed gel; the denaturant concentration represents the mid-point of the curve and is characteristic of the sequence. Introduction of a GC-rich sequence, which is very stable to denaturation and thus acts as a *GC clamp*, into the DNA allows detection of mutations in the more stable regions of the DNA.

dendrimer: A highly-branched, well-defined macromolecule that is usually composed of tiers of that radiate from a central unit. Each tier is composed of the same subunit, as it represents a discrete stage of the synthetic process.

dendrogram: A tree-like representation, e.g. as a phylogenetic tree, that shows derivations, with developments usually rising above or radiating from the origin located at the bottom or centre.

density-gradient centrifugation: A technique to separate and/or characterize a macromolecule by high-speed centrifugation in a density gradient formed by the concentration gradient of a solute such as CsCl or sucrose. The macromolecule sediments until it reaches the zone of its own density. Also known as isopycnic centrifugation or zonal centrifugation.

deoxy-: A prefix that signifies a product of the replacement of a hydroxy group by a hydrogen atom, e.g. deoxycorticosterone, deoxyribose.

deoxyribonucleic acid: (*see* **DNA**)

depletion-insertion: A technique used to define properties of a cellular, subcellular or tissue model, in which a variable or undesired constituent, e.g. membrane phospholipid, is removed (by a phospholipase) and then replaced by a substitute constituent of defined characteristics (e.g. phosphatidylserine).

depside: An ester of two or more derivatives of orsellinic acid (1,2-dihydroxy-5-methylbenzoic acid). A *depsidone* is a depside in which the aromatic subunits are additionally coupled to each other by an ether linkage. (*see also* acetate rule; acetogenin)

depsidone: (*see* depside)

depurination: The cleavage of N-glycosidic bonds of DNA to form apurinic DNA.

desensitization: The loss of responsiveness of an enzyme to allosteric regulation while retaining its catalytic activity.

designability: A hypothetical property of a protein folding motif; its relative thermodynamic and evolutionary stabilities, i.e. resistance to denaturation and changes in its primary sequence, respectively. A motif is designable, because it can be achieved by a variety of amino-acid sequences, and therefore can be encoded by a variety of DNA sequences. Designability encompasses foldability, a putative thermodynamic stability which is greater for any real motif than that for any other folded state that may be achieved by the same amino-acid sequence.

desmolase: An enzyme that oxidatively cleaves carbon-carbon bonds, e.g. the side-chain-cleaving enzyme that converts cholesterol into pregnenolone.

desmosome: A type of junction that attaches one cell to its neighbour.

desorption: (*see* **mass spectrometry (MS)**)

desoxo-: A prefix that signifies a product of the replacement of an aldehyde or ketone carbonyl oxygen with two hydrogen atoms.

desoxy-: (= **deoxy-**)

destruction box: A sequence of about 10 amino acid residues that determines a protein's susceptibility to ubiquitinization and subsequent proteolysis.

detergent (surfactant): An amphipathic compound able to stabilize suspensions of non-polar materials in aqueous solution.

detoxification: The chemical modification by oxidation, methylation, glycosylation, etc. of a xenobiotic to render it innocuous.

dextran: A branched-chain storage polysaccharide of microbial origin.

dextrin: (*see* **Schardinger dextrin**)

dextro-rotation: (*see* **optical rotation**)

DGGE technique: (*see* denaturing gradient gel electrophoresis).

di-: A prefix derived from the Greek for two; also the form of the prefix dia- used before a vowel. In organic chemistry, used to denote twice the usual proportion of an element or other component; bi- was previously used in the same sense.

dia-: A prefix derived from the Greek for through.

diagonal electrophoresis: A method for identification of a particular kind of peptide in a mixture by identical electrophoretic steps, the second at a 908 angle to the first, with a chemical modification introduced between the steps. An example is the identification of tyrosine-bearing peptides in a mixture of peptides by treatment of the products of paper electrophoresis with iodine vapour to iodinate the tyrosine residues before the second electrophoresis. Iodotyrosine peptides are then identified as those that deviate from the diagonal formed by all the other peptides when they are visualized by, for instance, the ninhydrin reaction.

dialysis: A technique for the separation of macromolecules from smaller molecules by placing them within a semi-permeable membrane, such as Cellophane, separating them from a large volume of water. Only the low-molecular-mass diffusible molecules cross the membrane and pass into the larger volume; the macromolecules are confined to their original space. *Equilibrium dialysis* is the technique of quantification of binding capacity and affinity by dialysis of a macromolecule against various concentrations of a ligand and subsequent measurement of the final concentrations of bound and free ligand within the dialysis chamber and free ligand outside it.

diamagnetic: Descriptive of a compound or chemical group (usually one that contains no unpaired electrons) that is not affected by a magnetic field. (*see also* paramagnetic)

diaphorase: An enzyme that transfers electrons from NADH to a dye or to ferricyanide.

diastereoisomer: One of two or more compounds that differ from each other at one or more asymmetrical centres, e.g. D-erythrose and D-threose. (*see also* enantiomer; epimer; stereoisomer)

diauxie: Bacterial growth in two stages which occurs when the bacteria are grown on two carbon sources, e.g. glucose and xylose. The first phase corresponds to consumption of one compound, followed by a lag phase until enzymes for the assimilation and metabolism of the second compound are produced.

Dickens-Warburg pathway: (= pentose phosphate pathway)

Dictyostelium discoideum: A cellular slime mould; the subject of much experimentation because it offers a mass of cells with synchronized cell cycles.

dideoxynucleotide fingerprinting (ddF): A method for the detection of mutation in a DNA fragment. The DNA serves as a template for replication using a mixture of four normal deoxynucleotides plus one dideoxyribonucleotide. Electrophoresis on a non-denaturing gel shows shifted mobility for the mutant template products compared with those of the normal DNA template. (*see also* single-strand conformational polymorphism (SSCP))

dideoxynucleotide sequencing: (= Sanger method (dideoxynucleotide sequencing))

dideoxynucleotide: A 2',3'-dideoxynucleoside 5'-triphosphate; a deoxynucleotide analogue that lacks a hydroxy group at its 3'-carbon and functions as a chain-terminator during DNA synthesis. (*see also* Sanger method (dideoxynucleotide sequencing))

difference cloning: A group of techniques (e.g. subtractive DNA cloning) designed to isolate DNA sequences that are not shared between two DNA sources, e.g. cDNA clones prepared from different tissues, or under different conditions. Difference cloning is distinguished from *coincidence cloning*, comprising techniques (e.g. end ligation coincident sequence cloning) designed to isolate DNA sequences that are shared between two DNA sources.

difference Fourier method: A technique of X-ray crystallography for solving the structure of a protein-ligand complex once the structure of the original protein has been deduced, by determination of the difference in distribution of densities and, eventually, the electron densities that result from altering a crystal of the protein by diffusing the ligand into it.

difference Patterson map: An interim stage in the solution of a molecular structure from X-ray crystallographic data. Using only the intensities due to heavy atom reflections obtained by subtraction of the reflections of the protein from those of an isomorphous replacement, Fourier analysis calculates the vectors between heavy atoms. Their display is the difference Patterson map. This map may then be used to calculate the relative positions of the heavy atoms. (*see also* Fourier transformation; phase problem)

differential centrifugation: The fractionation of subcellular components according to their sedimentation behaviour; separation into nuclei, mitochondria, lysosomes, microsomes (endoplasmic reticulum), ribosomes, cytosol, etc. by removal of sedimenting material after cycles of processing at progressively increasing centrifugal force.

differential chemical modification: (*see* competitive labelling)

differential genome display: A method for identifying candidate genes that may determine significant differences between closely related species whose genomes have been mapped, e.g. between a pathogen and a closely related non-pathogenic bacterium. The genome of one species is displayed as a circle showing at points on its circumference the genes. The genes are assigned a colour code according to a priority of similarities between the species: closely homologous genes are selected first, then the less closely related, then those of the remaining genes that have other known functions (viral proteins, bacterium–host interaction factors) until finally the unknown candidate genes remain. The display may also show a smaller concentric circles which display G/C content or other potentially relevant information.

differential hybridization: A method for comparison of different organs, or physiological or disease states by the genes that are uniquely expressed in them. The mRNAs of two types of cell to be compared, e.g. normal and cancer cells, are isolated, differently labelled for detection and hybridized to an appropriate cDNA library, preferably a normalized library. The library can be in the form of clones blotted onto duplicate filter membranes and the two sets of mRNAs, or can be labelled with two different radioisotopes, so that a clones that are equally labelled on both membranes identify a gene that is expressed equally in both kinds of cell, and a clone that fails to appear on one membrane will indicate transcription in only the other type of cell. A more sophisticated version labels the mRNAs with two different fluorophores, e.g. red emission for cancer cells and green emission for normal cells, so that when they are hybridized to the cDNA library that is arrayed in a relative excess on a single glass surface, the hybridization of both fluorophores to a single cDNA fragment will appear yellow, while the hybridization of only one, e.g. the normal cell mRNA, will appear green.

differential PCR display: A method for identification of mRNAs produced under specific physiological conditions. Total cellular RNA is reverse-transcribed, and the resultant cDNAs are used as templates for PCR. The 3'-primer has a poly(T) sequence that directs it to the poly(A) tail of mRNA, and is 5'-ended with two bases that make this primer more selective. (The two bases may be varied to achieve different selections.) The 5'-primer is short and of arbitrary sequence, and is intended to allow, under controlled non-stringent reannealing conditions of temperature, concentration, time, etc., amplification of a manageable number of specific cDNAs (between 50 and 100 bands) that can be electrophoretically separated on a DNA sequencing gel. The ladder of cDNAs can be displayed side-by-side with the cDNAs derived from cells in a different physiological state. This allows the identification of unique cDNAs that can be extracted, amplified by PCR, sequenced and identified. In a variant version called *RNA fingerprinting arbitrary primer PCR* (*RAP-PCR*), no poly(T) primer is used; one primer can serve for each of the two transcribed strands and the non-stringent reannealing conditions allow a fit to a manageable number of sites on the cDNA template.

differential reassociation: (*see* subtractive DNA cloning)

differentially methylated region (DAR): The process by which some genes are rendered non-equivalent. The paternal or maternal allele is not expressed (allelic exclusion), or is expressed differently in different tissues. The phenomenon is possibly directed by an imprinting box, a parent-specific polynucleotide sequence that instructs an imprinting factor, which may act by methlyation of differentially methylated regions (DMRs), and which will reversibly activate or inactivate the gene inherited from only one parent. Imprinting passes through erasure of methylation of both parental and maternal chromosomes during germ cell development, establishment of methylation, according to the sex of the gamete, and after fertilization, maintenance of the methylation status of the genome during the mitoses of embryogenesis.

diffraction pattern: The array of reflections obtained by crystallography. Each reflection indicates an intensity and, by its location, the angle with respect to the incident beam. Because a sharp pattern requires the wavelength of the incident beam to be of the same order of magnitude as the regular spacings, determination of inter-atomic distances is performed by X-ray crystallography, and the X-ray diffraction pattern is used to construct a three-dimensional model of the crystal. (*see also* subtractive DNA cloning)

diffusion coefficient: A measure of a molecule's ability to travel through a solution, propelled by Brownian movement; expressed as RT/Nf ($cm^2 \cdot s^{-1}$), where R is the gas constant, T is the absolute temperature, N is Avogadro's number and f is the frictional coefficient.

diffusion limited: Descriptive of a rate, e.g. of an enzymic reaction, that is determined only by the rate of diffusion of reactants towards the enzyme or of products away from it.

digestibility: The characteristic of a protein that accounts for its nutritional quality according to how efficiently its digestion delivers its amino acids, especially the essential amino acids, to the individual.

digestion: The degradative action of hydrolytic enzymes or of reagents such as acids.

dihydro-: A prefix that signifies a product obtained by the addition of two hydrogen atoms to a precursor.

di-iron: Descriptive of metalloproteins in which two iron ions are bridged by two carboxylate groups or by a carboxylate group and an oxide.

diketopiperazine: A six-membered heterocyclic product of the condensation of two α-amino acids that contains amide bonds between the two α-amino and the two α-carboxy groups.

dimedone: 5,5-Dimethyl-1,3-cylcohexanedione; a reagent that has been used to assist in the isolation of aldehydes and in their quantification (e.g. for [^{14}C]formaldehyde by its radioactivity) because it forms a water-insoluble, crystallizable adduct.

Dintzis experiment: The method by which it was demonstrated that proteins are synthesized in the N- to C-terminal direction. A labelled amino acid was supplied to a reticulocyte preparation synthesizing haemoglobin and the completed globin chains were isolated after a brief incubation. The radioactive products were found to be concentrated in the C-terminus, indicating that this was the last part of the molecule to be formed.

dioxygenase: An enzyme that reduces molecular oxygen by incorporating both atoms into its substrate, e.g. tryptophan dioxygenase.

diploid: Descriptive of the number of chromosomes of a somatic cell, i.e. two of each chromatid. (*see also* haploid)

direct bilirubin: (*see* van den Bergh reaction)

direct calorimetry: Evaluation of the heat evolved by a human or experimental animal by measurement of the heat exchanged with the environment in specially constructed insulated chambers. (*see also* indirect calorimetry)

direct van den Bergh reaction: (*see* van den Bergh reaction)

directed proton transfer: (*see* entropy effect)

dirigent: A protein, itself devoid of catalytic activity, taht directs the regio- and stereo-selective course of an enzymic reaction, e.g. the exclusive synthesis by an oxidase of (+)-pinoresinal from two E-coniferyl alcohol molecules.

disc-gel electrophoresis: A technique for electrophoresis of single samples in an open-ended tube. The polyacrylamide stationary phase is polymerized *in situ*. During electrophoresis the ends of the tube are immersed in upper and lower buffer chambers through which it is connected to the power supply. The highly focused gel travels through the tube as a disc. (*see also* slab gel)

discontinuous transcription: (*see* trans-splicing)

disease gene mapping: The localization of a specific gene to a locus on a chromosome by study of the morphology of chromosomes and by linkage studies.

dispersion optics: (*see* Fourier transform infrared spectroscopy)

displacement chromatography: A column chromatographic method for separation and concentration of components of a solution. After adsorption of the sample to be separated on to the resin in a column, all bound molecules are displaced by a concentrated solution of a displacer, whose affinity for the stationary phase must be higher than that of any of the sample molecules. As the displacer sweeps the adsorbed molecules ahead of itself, each of these becomes a displacer for any molecule of lower affinity. Thus the original components are eluted in small volumes ahead of the displacer, in reverse order with regard to their affinities for the stationary phase.

disrupter: (*see* endocrine disrupter)

dissociation constant (K_d): Given by [A][B]/[AB], where [AB]=[A]+[B]; expressed in units of concentration. (*see also* association constant (K_a); pK_a)

distributive control: The principle that the control of flux of metabolites through a pathway resides in several enzymic steps and not, as proposed by simpler models, in a single rate-controlling step.

disulphide bridge: An inter- or intra-polypeptide cross-link formed by oxidation of the thiol groups of two cysteine residues to a single cystine residue.

diterpene: (*see* terpene)

Dixon plot: A graphical method for determination of the type of enzyme inhibition and the dissociation constant (K_i) for an enzyme-inhibitor complex. The effect on the enzymic rate (v) is determined at two or more substrate concentrations, and over a range of inhibitor concentrations (*I*). In a plot of $1/v$ against *I*, data for each substrate concentration fall on straight lines that intersect at I=-K_i and $1/v$=$1/V_{max}$ (competitive inhibition), or that intersect on the abscissa ($1/v$=0) at I=-K_i (non-competitive inhibition). For uncompetitive inhibition, the lines are parallel. (*see also* Cornish-Bowden plot; Easson and Stedman plot; Hunter and Downs plot)

DNA: Deoxyribonucleic acid; a macromolecule formed of repeating deoxyriboses linked by phosphodiester bonds between the 3-hydroxyl group of one and the 5-hydroxyl group of the next. A purine, adenine or guanine, or a pyrimidine, cytidine or thymine, is held in a glycosidic bond to the anomeric carbon of the sugar. DNA appears in Nature in both double-stranded (the Watson-Crick model) and single-stranded forms, and functions as a repository of genetic information that is encoded in its base sequence. (*see also* B-DNA; Z-DNA)

DNA chip: *see* chip (oligonucleotide array)

DNA enzyme: A single-stranded DNA that can hydrolyse a complementary RNA sequence. Such activity has been observed only in some synthetic oligodeoxynucleotides.

DNA fingerprinting (DNA profiling): A method to generate a pattern of DNA restriction fragments that is unique to an individual, especially for forensic purposes. DNA from blood, semen or another tissue sample is isolated and cleaved with a restriction endonuclease; the products are separated by polyacrylamide-gel electrophoresis, blotted on to a nylon sheet, and fragments that contain a specific nucleotide sequence are detected by hybridization with an appropriate DNA probe to produce the unique pattern. (*see also* variable number tandem repeat (VNTR))

DNA gyrase: An enzyme that uses the energy of ATP hydrolysis to unwind double-stranded circular DNA to form a negatively supercoiled molecule.

DNA methylation: A phenomenon that represses expression of regions of the genome. Transcription is prevented when the DNA is methylated and folded into nucleosomes. Eukaryotic DNA is methylated almost exclusively as 5-methyl cytosine; prokaryotic

DNA is methylated also as 6-methyl adenosine. (*see also* imprinting)

DNA molecular decoy: A synthetic polynucleotide with high affinity for a regulatory protein, such as a transciption factor, that can be used for research purposes, and potentially for therapy, to compete with the natural DNA sequence and attenuate the effects of the regulatory protein.

DNA profiling: (= DNA fingerprinting (DNA profiling))

DNA repair: The removal of damaged segments, e.g. pyrimidine dimers, from one strand of double-stranded DNA and its correct resynthesis.

DNA typing: (= DNA fingerprinting (DNA profiling))

DNase footprinting: (= exonuclease footprinting (DNase footprinting))

dNTP: Deoxynucleoside triphosphate; the letter N refers to any or all of the common bases.

docking protein: (*see* signal recognition particle (SRP))

docking: In a computer graphics simulation, the bringing together of two molecular models to explore their interactions, e.g. the insertion of a substrate into the active site of an enzyme.

dogma: (*see* central dogma)

domain boundary model: (*see* insulator)

domain swapping: The formation of a protein homo-oligomer by the replacement of a domain of each protomer with the same domain of an adjacent protomer.

domain: A region of a protein that has a stable secondary and tertiary structure that can fold independently of other domains; often connected to other domains, of the same protein if any, by short sequences without secondary structure. Domains are typically encoded by a single exon, and boundaries between domains are marked by introns. (*see also* exon shuffling; mosaic protein; motif)

dominant: (*see* allele)

Donnan effect (Gibbs-Donnan effect): The unequal distribution of a diffusible ion across a semi-permeable membrane when an impermeable electrolyte, such as a protein, is also present on one side; e.g. a solution of an anionic protein with Na^+ as a counterion, in contact through a membrane with a NaCl solution, will result in the transfer of Cl^- into the protein compartment and an equivalent amount of Na^+ out of it, such that the products of concentrations, $[Na^+]3[Cl^-]$, in each compartment are equal.

donor (5') splice site: The site on pre-mRNA that corresponds to the 3'-end of an exon and the 5'-end of an intron. The first two bases of the intron at the donor splice site are GU in most cases, but AU in some instances.

donor: (*see* acceptor)

DOP: Degenerate oligonucleotide primed PCR. (*see* coincidence painting (chromosome painting))

dot blotting: A method to estimate the concentration of a polynucleotide or protein solution by spotting it on a sheet of nitrocellulose or nylon, hybridizing it with a radiolabelled complementary polynucleotide or antibody, visualizing it by autoradiography, then comparing its intensity with that of a similarly treated graded concentration series of the standard polynucleotide or protein.

double adaptor method: cDNA concatamer sequencing; a method for increasing the efficiency of sequencing cDNA library subclones cDNA fragments are cleaved with a restriction nuclease and randomly religated into larger DNA fragments (mis-named concatamers), which are then randomly sheared and the ends, repaired. A 5'-overhanging adaptor is ligated to them, which allows their insertion into a vector which has been prepared by leavage and religation with complementary oligonucleotide 5'- overhangs. The vectors with inserts are used for transfection. When these inserts are sequenced, the artificial junction between religated fragments may be recognized by the presence of the restriction site. The method increases the efficiency of sequencing of what would otherwise be 3'-ends of fragments, which are less likely to be successfully sequenced.

double duplex invasion: A mechanism by which a pair of pseudocomplementary protein nucleic acids (PNAs) target and inactivate DNA. The DNA strands are unwound locallly and each is targeted simultaneously by one of the PNAs.

double helix (Watson-Crick model): The arrangement in space of two polynucleotide chains in which each chain is wrapped around the other to form two antiparallel spirals. Each strand presents to the other the bases, purine to pyrimidine, with which it can form inter-strand hydrogen bonds. (*see also* A-DNA; Z-DNA)

double minute chromosome (DMC): An autonomously replicating structure found, for example, in leukaemic cells.

double sieve: Descriptive of the mechanism for assuring correct charging of a tRNA by its amino acid activating enzyme. Only an amino acid that exactly fits the binding site will be ligated to the nucleic acid; those that are too large do not bind to the active site and those that are too small are hydrolysed. (*see also* proofreading (editing))

double-displacement kinetics: (= Ping Pong Bi Bi kinetics; *see* enzyme mechanism)

double-reciprocal plot: (= Lineweaver-Burk plot)

double-sieve editing: A proposed mechanism that accounts for the high selectivity of some enzymes, especially the amino acid activating enzymes. An initial screening sorts substrates by size - those too large cannot enter the binding site - while those smaller than optimal may be activated by immediate hydrolysis *in situ*. (*see* kinetic proofreading)

double-strand-break repair model: double-strand-break repair model A mechanism proposed for the repair of DNA that has been fragmented at in a non-homologous region. The strands are unwound at the two ends to expose a site of microhomology, a potential pairing of no more than ten bases of each of the two strands. The strands are placed together at the site of microhomology, unpaired bases are cleaved, missing bases are filled in and ligated. Protein factors act to hold together the ends and to catalyze the necessary reactions. The process is closely analogous to the formation of a coding joint in V(D)J recombination of immunoglobulin synthesis.

double-stranded: Descriptive of two complementary polynucleotide strands paired in a double helix; also (less commonly) of a protein multimer, such as F-actin.

doublet: A subunit of the axoneme in which 10 microtubule protofilaments are grafted longitudinally on to a whole microtubule, which itself consists of 13 protofilaments. Also, in magnetic resonance, a peak that has been split into two by the interaction of a resonance with a nearby perturbing centre. (*see also* splitting)

double-well hydrogen bond: (*see* hydrogen bond)

downfield: Descriptive of a resonance position in a magnetic field lower than that at which a standard displays its signal.

down-regulation: The decrease in the number of hormone receptors on a target cell that occurs after exposure to the hormone. Also, in bacterial nutrition, equivalent to down-shift.

down-shift: The response by cells, especially micro-organisms, to an altered nutritional state, e.g. the decrease in ribosomes or enzyme production when levels of some nutrients are low.

downstream: In a polynucleotide chain, towards the 3' end. (*see also* upstream)

DPN^+: Diphosphopyridine nucleotide; obsolete name for NAD^+.

DPNH: Reduced diphosphopyridine nucleotide; obsolete name for NADH.

driver: (*see* subtractive DNA cloning)

Drosophila melanogaster: The fruit fly; the subject of classical genetic studies, because of its rapid generation time and easily observed characteristics

dsDNA: Double-stranded DNA. (*see also* double helix (Watson-Crick model))

dsRNA: Double-stranded RNA (*see* RNA interference)

dual-recognition model: (*see* altered-self hypothesis)

duplex: A double-stranded polynucleotide.

dUTP system mutagenesis: (= Kunkel mutagenesis (dUTP system mutagenesis))

duty cycle: (*see* molecular motor)

dyad symmetry: A property of double-stranded DNA whereby one strand has the same base sequence (5'-to-3') as its complementary strand (5'-to-3'), which allows it to bind to a homodimeric or homotetrameric binding protein, e.g. the *lac* repressor. In practice, the term is applied to polynucleotide sequences that are more extensive, but have less perfect symmetry, than the palindromic sequences that are sites of restriction nuclease cleavage.

dye primer: (*see* Sanger method (dideoxynucleotide sequencing))

dye terminator: (*see* Sanger method (dideoxynucleotide sequencing))

E

E. coli: (= Escherichia coli (E. coli))

Eadie-Hofstee plot: A linearized display of kinetic data of dependence of enzyme activity on substrate concentration; rate is plotted against the ratio of rate to substrate concentration. (*see also* Lineweaver-Burk plot)

Easson and Stedman plot: A graphical method for determination of enzyme-inhibitor dissociation constants for non-competitive inhibitors, especially those that bind very tightly. Where I is the total inhibitor concentration, E is the total enzyme concentration and α is the ratio of inhibited to non-inhibited rates under the same conditions, I/α is plotted on the ordinate (y-axis) against $1/(I-\alpha)$ on the abscissa (x-axis) to give a straight line that intersects the ordinate at E; its slope is K_d, the dissociation constant. (*see also* Cornish-Bowden plot; Dixon plot; Hunter and Downs plot)

EC_{50}: The concentration of an effector that evokes a response in 50% of the subjects, animals, cells, etc. in which it is tested.

eccrine secretion: (= exocytosis)

ecto-: A prefix derived from the Greek for outside. *see* endo-

ectoderm: One of the three primordial germ layers formed during early embryogenesis; a precursor of the central nervous system, sensory organs, adrenal medulla, etc. (*see also* endoderm; mesoderm)

ectodomain: *see* membrane protein secretase

ectoenzyme: An enzyme located on the outer surface of the plasma membrane with its active site directed to the extracellular space so as to be able to catalyze reactions in the extracellular space; e.g. 5'-nucleotidase.

ED_{50}: Like EC_{50}, the dosage of a substance that achieves an effect in 50% of the tested individuals.

editing: In DNA replication, the 3'-to-5' exonuclease activity of a polymerase that removes incorrectly paired bases, also referred to as proofreading; also, loosely applied to some of the changes during processing of pre-mRNA, e.g. deamination of some adenylate residues to form inosinyl residues.

editosome: A hypothetical multienzyme complex which in trypanosomes performs RNA editing, especially the deletion and insertion of U residues.

Edman degradation: A chemical technique to degrade and cleave amino acid residues sequentially from a protein beginning at the N-terminus, and to identify the residues as they are removed. Reaction of the N-terminal residue with phenylisothiocyanate cleaves it from the protein as the phenylthiohydantoin derivative, which may be isolated and identified. (*see also* thiocyanate degradation)

EF hand: The region in some homologous calcium-binding proteins where the ion is located. The E and F helices of calmodulin, troponin C, etc., resemble the extended index finger (E helix) and thumb (F helix) of the right hand. The ion is located in the pocket formed by the closest approach of the E and F helices to each other.

effector: A compound that modulates an allosteric enzyme; a system that produces an intracellular response to a hormone, e.g. adenylate cyclase.

efferent: *see* afferent

eicosanoid: One of a class of compounds that includes the prostaglandins, thromboxanes and leukotrienes, all derived from arachidonic (eicosatetraenoic) acid.

Eisenberg plot: A graphical representation of the hydrophobic moment; μ_M; against the mean hydrophobicity ;H; of an α-helix. α-Helices of globular proteins generally have low; μ_M; and low; H; values; membrane surface-associated helices have high; μ_M; and high; H; values; and α-helices imbedded in membranes have low; μ_M; and high; H; values.

EL-CSC: (= end ligation coincident sequence cloning (EL-CSC))

electric field effect: (*see* entropy effect)

electrical breakdown: (= electropermeabilization)

electroblotting: An electrophoretic technique for transfer of a protein or nucleic acid from a slab gel, or similar unstable two-dimensional matrix, to a sheet of stable material such as nitrocellulose or poly(vinylidene difluoride), for further operations or analysis. The electric field is perpendicular to the plane of the gel and the sheet. (*see also* electroelution)

electrochemical gradient: A concentration and charge difference across a membrane, e.g. the pH difference and membrane potential developed during mitochondrial electron transport. (*see also* chemiosmotic theory)

electroelution: A technique to remove a sample previously purified by electrophoresis on a solid support, by electrophoresing it into a buffer. (*see also* electroblotting)

electrofuge: A chemical group that is displaced in an electrophilic substitution or elimination reaction. (*see also* nucleofuge)

electromicrofiltration: A separation procedure in which an electric field is imposed upon an ultrafiltration membrane in order to prevent non-permeable material from accumulating on and clogging it.

electron density map: In X-ray crystallography, a representation that resembles a geological survey map of planes drawn through a crystal. It shows contours of equal electron density surrounding the individual atoms: steeper around heavier atoms; more shallow around lighter atoms.

electron nuclear double resonance (ENDOR): A technique for detection of coupling between electrons and nuclei in order to gain information about the valance electron distribution around a paramagnetic nucleus; observation of an electron spin resonance transition while applying radio frequency energy to effect nuclear magnetic resonance transitions.

electron paramagnetic resonance: (= electron spin resonance (ESR))

electron sink: In discussions of enzyme mechanisms, a group that can pull electrons from a reactive centre and thus stabilize an electron-deficient intermediate or transition state, e.g. pyridoxal phosphate condensed with an a-amino acid.

electron spin resonance (ESR): Also known as *electron paramagnetic resonance (EPR)*; a magnetic resonance technique that detects and characterizes the environment of a moiety that contains an unpaired electron by the energy that is equivalent to a change in its spin state. (*see also* spin label)

electron tomography: A technique for visualizing subcellular structures at very high resolution (near 2.5 nm). The material is prepared for electron microscopy; electron micrographs are taken as the sample is tilted + or - 60° in small steps (e.g. 2°). The data are then computer-manipulated to recreate images of the structure in three-dimensional space. Also, electron microscopic images of random images of a structure may be reconstructed into a three-dimensional image of somewhat lower resolution.

electron transport: The mediation of oxidation of one metabolite and the reduction of another by a series of carriers, cytochromes, iron-sulphur proteins, quinones, etc.

electron transport chain (respiratory chain): Usually, mitochondrial electron transport.

electron transport particle: (= submitochondrial particle (SMP))

electronegativity: The degree to which an atom or chemical group holds its electrons in competition with the atoms or groups to which it is bonded.

electronic film: (= area detector)

electronic PCR (e-PCR): *see* sequence-tagged site

electronic strain: The distorted distribution of electrons in a substrate induced by the proximity of a strong dipole in the enzyme, e.g. Zn^{2+} in carboxypeptidase.

electron-transfer potential: A quantitative measure of the ability of a molecule to lose an electron in a redox reaction under standard conditions; expressed in volts (V). (*see also* redox potential (Eo'))

electro-osmosis: *see* iontophoresis

electropermeabilization: A technique for the introduction of materials into an intact cell. Cells are subjected to a strong electric field which causes them to become temporarily permeable and thus take up molecules from their environment that would otherwise remain excluded from them. Also known as electrical breakdown, electropermeation, electroporation and high voltage electrical discharge.

electropermeation: (= electropermeabilization)

electropherogram: A graphical representation of an electrophoretic separation; e.g. for a slab gel, absorbance (ordinate) plotted as a function of distance from the origin (abscissa)

electrophile: A compound or functional group that can attract an electron pair. (*see also* nucleophile)

electrophoresis: A technique that separates charged compounds according to their mobilities in an electric field. In *free zone electrophoresis* the compounds are in a solution unsupported by any stabilizing matrix. In *gel electrophoresis* the solution is stabilized by a gel (e.g. agarose, polyacrylamide, starch). In *paper electrophoresis* the stabilizing matrix is buffer-soaked paper. In *isotachophoresis*, by contrast, charged solute molecules all travel at the same rate: a highly mobile leading ion (e.g. Cl^-) travels first and a relatively immobile ion (e.g. tricine) is the trailing ion; to avoid an electrostatic gap, other solute molecules travel between them, all at the same rate but ordered according to their relative mobilities. *Capillary eletrophoresis* uses a long, very-narrow-bore tube coated with the stationary phase; as the mixture passes along the length of the lumen, components are differentially retarded to effect a separation. (*see also* rocket electrophoresis)

electrophoretic mobility shift assay (EMSA): (= gel shift assay (electrophoretic mobility shift assay; EMSA))

electroporation: (= electropermeabilization)

electrospray mass spectrometry (ES-MS): (*see* mass spectrometry (MS))

electrostatic stabilization: (*see* entropy effect)

electrostriction: The contraction of a solution volume in the presence of ionizing solutes, due to the more compact organization of water molecules around the charged groups.

ELISA: Enzyme-linked immunosorbent assay; a technique that combines the specificity of an immunoglobulin with the detectability of an enzyme-generated chromophoric product to quantify a macromolecule. The enzyme is covalently attached to the immunoglobulin; when the latter is adsorbed to an antigen, its presence is revealed by the enzymic generation of a chromophore.

elongation cycle: The reactions of a ribosome that add one amino acid residue to the C-terminus of a growing polypeptide chain and move the ribosome three nucleotides towards the 3'-end of the mRNA.

elongation factor: An accessory protein that is required for transfer of amino acid residues to the polypeptide chain during translation.

elution: In chromatography, the washing out of an adsorbed material from a solid support, especially by conditions that assist displacement, e.g. salt concentration, altered pH.

elutriation centrifugation: (= centrifugal elutriation)

Embden-Myerhof pathway: The sequence of enzymic reactions that convert glucose into pyruvic acid; also called *aerobic glycolysis*. (*see also* glycolysis)

-emia: (*see* -aemia)

emission spectrum: In fluorescence spectroscopy, emission as a function of wavelength, upon excitation at a fixed, shorter wavelength. (*see also* excitation spectrum)

EMSA: Electrophoretic mobility shift assay (= gel shift assay (electrophoretic mobility shift assay; EMSA))

enantiomer: The mirror image of an asymmetrical compound. (*see also* diastereoisomer; epimer; stereoisomer)

enantioselective: Descriptive of a technique that results in one enantiomer in preference to the other. Enzymic reactions are typically enantiospecific, i.e. produce only one of the possible enantiomer products.

end ligation coincident sequence cloning (EL-CSC): A technique for selection of DNA sequences common to two DNA populations. Of the two populations, the one of lowest complexity (or one made least complex by the creation of representations) is used in excess over the other population; it is cleaved by a restriction endonuclease and the fragments are ligated to linker oligonucleotides that will serve as primers, one of which is biotinylated; the other DNA population is cleaved by the same endonuclease. The two populations are mixed, denatured and annealed, and biotinylated fragments (i.e. including those from the second population that are also represented in the first) are isolated on a solid support, based on the presence of biotin. (The non-shared fragments from the second source are washed away.) The recovered fragments are then exposed to 'capture' oligonucleotides that are complementary to the linker oligonucleotides and will therefore be placed adjacent to the strands that originated in the second population; these strands are then ligated to the capture oligonucleotides. PCR using primers complementary to the capture oligonucleotides amplifies these fragments, which can be released from the primers by the endonuclease.

endergonic: Descriptive of a chemical reaction that consumes free energy. (*see also* endothermic; exergonic)

end-group analysis: A technique to determine the C-terminal or N-terminal residue of a protein. (*see also* Edman degradation; thiocyanate degradation)

endo-: A prefix derived from the Greek for within.(*see also* exo-; ecto-)

endocrine: Descriptive of secretion into the bloodstream, e.g. insulin secretion. (*see also* autocrine; exocrine; paracrine)

endocrine disrupter: An environmental chemical, e.g. an herbicide, that unintentionally mimics or interrupts the normal action of a hormone and thus causes medical or biological problems in humans or animals.

endocytosis: A process by which extracellular matter is taken up by a cell; the plasma membrane invaginates and encloses the material in newly formed vesicles. In *phagocytosis* the extracellular material is particulate; in *pinocytosis* it is non-particulate. (*see also* exocytosis)

endoderm: One of the three primordial germ layers formed during early embryogenesis; a precursor of the gastrointestinal system, digestive glands, liver, lungs, etc. (*see also* ectoderm; mesoderm)

endo-form: (*see* envelope conformation)

endogenous: Arising within the organism or cell. (*see also* exogenous)

endopeptidase: A peptidase that cleaves a protein at an internal peptide bond. (*see also* exopeptidase)

endoplasmic reticulum (ER): An extension of the nuclear envelope that forms sheets of membranes that are generally parallel to the nucleus of a cell. (*see also* rough endoplasmic reticulum (RER))

endoplasmic reticulum-associated protein degradation (ERAD): The transport of unfolded or mis-folded proteins from the endoplasmic reticulum to the cytoplasm, where they are degraded by proteasomes. *see* unfolded protein response

ENDOR: (= electron nuclear double resonance (ENDOR))

endorphin: Endogenous morphine-like peptide; one of a class of peptides that are derived from pro-opiomelanocortin by limited proteolysis and have analgaesic properties. The group includes the 31-residue β-endorphin, and two pentapeptides, Leu-enkephalin and Met-enkephalin. (*see also* enkephalin)

endosome: (= receptosome (endosome))

endosymbiotic hypothesis: The proposal that some specialized organelles of contemporary organisms, e.g. chloroplasts and mitochondria, arose during evolution from the invasion of a eukaryotic cell by a bacterium.

endothermic: Descriptive of a chemical reaction that consumes heat (*see also* endergonic; exothermic).

endotoxin: (*see* exotoxin)

end-point assay: An enzyme-based assay that measures the amount of material by the quantity of a substrate consumed or product formed over the course of a reaction. (*see also* kinetic assay)

energy charge: (= adenylate energy charge)

energy landscape: *see* funnel concept

energy minimization: (*see* molecular dynamics)

energy spectrum: A profile of the energy of the emissions from a radioactive atom; the fraction of total emissions from the decay of a radioactive atom, e.g. the β-emissions of ^{14}C as a function of the energy of those emissions.

energy transfer: (*see* resonance energy transfer)

enhanceosome: A highly-ordered cluster of tissue-specific transcription factors that assemble at regulatory elements of a eukaryotic gene and which, by their arrangement in space, act synergistically and purportedly account for the diversity and tissue specificity of gene expression.

enhancer: A 50 to 1500 bp dsDNA segment that up-regulates transcription of a gene. The enhancer may be up- or downstream or even within an intron of the gene. It is distinct from the promoter; its action may be limited to a specific cell type or developmental stage. (*see also* promoter)

enhancing screen: A device to enhance the sensitivity of autoradiography; the surface that is to be visualized is sandwiched between a photographic film and the screen, which is a film impregnated with a phosphor that is excited by the sample's γ-rays and which can enhance the primary emission from the sample with its own secondary emission.

enkephalin: A pentapeptide isolated from the brain which has opiate properties, e.g. Met-enkephalin (Tyr-Gly-Gly-Phe-Met), Leu-enkephalin (Tyr-Gly-Gly-Phe-Leu). (*see also* endorphin)

enol: The tautomeric form of a -CH_2-CO- group; -CH=COH-. In nucleic acid chemistry, the alternative representations of the purine and pyrimidine rings as fully aromatic, with the substituent oxygens as phenolic hydroxy groups (enols), or as parts of carbonyl groups.

enrichment: (*see* mole percent excess)

ensemble: *see* funnel concept

enterohepatic circulation: The movement of a metabolite, e.g. a bile acid, from secretion by the gall bladder into the lumen of the intestine, its absorption, transport to the liver, and re-secretion in the bile.

enthalpy: A measure of the internal energy of a system, comprising binding forces, pressure, etc., expressed as J/mol (or cal/mol). (*see also* second law of thermodynamics)

Entner-Doudoroff fermentation: The metabolism of glucose, e.g. by *Pseudomonas fluorescens*, via 6-phosphogluconic acid and 2-oxo-3-deoxy-6-phosphogluconic acid to pyruvate and glyceraldehyde 3-phosphate.

entropy: A measure of the disorder or randomness of a system, expressed in J/mol per degree K (or cal/mol per degree K). (*see also* second law of thermodynamics)

entropy effect: The acceleration of a reaction that occurs when reactive groups are constrained in a productive orientation, either intermolecularly, as on an enzyme surface, or intramolecularly as in a model compound. Depending upon which aspect of this phenomenon is being emphasized, the physical bringing together of the components or some special feature of their association, it has been referred to variously as antichimaeric assistance, approximation, catalytic configuration, the Circe effect, coupling between conformational fluctuations, directed proton transfer, electric field effect, electrostatic stabilization, FARCE (freezing at the reactive centres of enzymes), gas-phase analogy, group transfer hydration, oribital steering, propinquity effect, proximity effect, rotamer distribution, stereoelectronic control, stereopopulation control, substrate anchoring, togetherness, torsional stress and virbrational activation.

envelope conformation: A conformation of a five-membered ring, e.g. a furanose, in which four ring atoms lie in a plane and C-2 or C-3 (2-*endo* or 3-*endo*) is out of the plane. (*see also* twist conformation)

enzyme: One of a class of biological catalysts that is composed principally of a globular protein of one or more polypeptide chains. In some cases enzymes include covalently bound or tightly associated metal ions, prosthetic groups or carbohydrates; they range in molecular mass from around 10000 to several hundred thousand Da.

enzyme mechanism: An elaborate system of nomenclature has been proposed to describe, from the perspective of kinetics, the variety of enzyme mechanisms. *Uni*, *Bi*, *Ter* and *Quad* refer to the number of substrates or products (e.g. Bi Bi refers to an enzyme that is bi-reactant in both directions, and uni bi to an enzyme that is uni-reactant in the forward direction and bi-reactant in the reverse direction). *Ordered* describes a mechanism that has an obligatory sequence of addition of reactants (and dissociation of products), *sequential* a mechanism in which all reactants are added before any product is released, and *random* a mechanism where there is no obligatory sequence. *Ping Pong* describes cases in which one or more products are released before all of the substrates have added to the enzyme. *Iso* refers to situations in which the enzyme undergoes isomerization to another stable form during the catalytic cycle.

enzyme-activated inhibitor: (= suicide inhibitor; *see* mechanism-based inhibitor)

enzyme-linked immunosorbent assay: (= ELISA)

enzyme-substrate complex (Michaelis complex): The association of a substrate with an enzyme that is an obligatory intermediate in conversion of the substrate into the product of the enzymic reaction.

enzymic inverse PCR (EI-PCR): A method for site-directed mutagenesis which generates highly efficient plasmids and which uses only two PCR primers. The primers incorporate the same site for a restriction nuclease, the recognition site of which is 5' to the cut; the linear PCR product treated with such a nuclease, therefore, does not contain the recognition site, but leaves overhangs which facilitate re-cyclization. The interior of the plasmid contains the desired mutation sequence, and the sequence of the overhang may be designed to be the native sequence.

eocyte: An extremely thermophilic sulphur-dependent archaebacterium. (*see also* eukaryote)

eosinophil: A polymorphonuclear leucocyte containing granules that react with eosin, a histological stain for acidic substances.

epi-: A prefix derived from the Greek for above.

epigenetic state: The condition of a cell that has been programmed during early embryogenesis, e.g. into ectoderm, endoderm or mesoderm, to a developmental fate that will be expressed many generations later.

epigenetic: Literally, outside of the germ line; therefore characteristics passed from cells to progeny without altering germ line DNA, e.g. genetic imprinting and tissue-specific development in eukaryotes and genetic transmission of plasmids in prokaryotes.

epimer: A compound that differs from another by its configuration at only one asymmetrical centre. (*see also* diastereoisomer; enantiomer; stereoisomer)

episome: A polynucleotide that can be transferred from one bacterium to another, either as a discrete structure or as a sequence incorporated into the host genome. (*see also* plasmid)

epithelium: A sheet-like tissue that lines body vessels, cavities and surfaces, and can differentiate into secretory glands.

epitope: The antigenic determinant; i.e. that part of an antigen to which an antibody is directed; it defines the specificity of the antigen and consequently binds to the receptor on a T-cell or to the antibody. (*see also* paratope)

epitope library: (= random peptide library)

EPR: Electron paramagnetic resonance. (= electron spin resonance (ESR))

equatorial: (*see* axial)

equilibrium dialysis: (*see* dialysis)

equilibrium labelling: In metabolic studies, the exposure of a biological system (e.g. a tissue slice, a cell culture, mitochondria) to a radiolabelled metabolite over a relatively long time so as to establish a constant steady-state specific radioactivity in downstream metabolites, in order to observe the metabolic fate of individual atoms of the original metabolite. (*see also* pulse labelling)

equilibrium sedimentation ultracentrifugation: A technique for evaluation of the molecular mass of a polymer by determination of the extent of sedimentation at which the force exerted upon a macromolecule in a centrifuge is balanced by its diffusion due to Brownian motion. (*see also* sedimentation-velocity ultracentrifugation)

equine: Derived from the horse.

ER: (= endoplasmic reticulum (ER))

erasure: The process by which some genes are rendered non-equivalent. The paternal or maternal allele is not expressed (allelic exclusion), or is expressed differently in different tissues. The phenomenon is possibly directed by an imprinting box, a parent-specific polynucleotide sequence that instructs an imprinting factor, which may act by methlyation of differentially methylated regions (DMRs), and which will reversibly activate or inactivate the gene inherited from only one parent. Imprinting passes through erasure of methylation of both parental and maternal chromosomes during germ cell development, establishment of methylation, according to the sex of the gamete, and after fertilization, maintenance of the methylation status of the genome during the mitoses of embryogenesis.

ergotypic: (*see* anti-ergotypic)

ERK: Extracellular-signal-regulated protein kinase (= MAP kinase)

erythrocyte: An enucleated, mature, red blood cell.

Escherichia coli (E. coli): An enteric bacterium that inhabits the human intestine; much used for experimentation.

ESI: Electrospray ionization mass spectrometry. (= ES-MS; *see* mass spectrometry (MS))

ES-MS: Electrospray mass spectrometry. (*see* mass spectrometry (MS))

ESR: (= electron spin resonance (ESR))

essential amino acid: An amino acid that is not synthesized by an organism at an adequate rate (or at all) from other amino acids or metabolites; therefore one that is a dietary requirement.

essential fatty acid: A fatty acid that is a dietary necessity, e.g. the polyunsaturated fatty acids linoleic acid and linolenic acid.

EST: (= expressed sequence tag (EST))

establishment: The process by which some genes are rendered non-equivalent. The paternal or maternal allele is not expressed (allelic exclusion), or is expressed differently in different tissues. The phenomenon is possibly directed by an imprinting box, a parent-specific polynucleotide sequence that instructs an imprinting factor, which may act by methlyation of differentially methylated regions (DMRs), and which will reversibly activate or inactivate the gene inherited from only one parent. Imprinting passes through erasure of methylation of both parental and maternal chromosomes during germ cell development, establishment of methylation, according to the sex of the gamete, and after fertilization, maintenance of the methylation status of the genome during the mitoses of embryogenesis.

esterase: An enzyme that hydrolyses esters; unless otherwise indicated, esters of carboxylic acids. The category includes the lipases, which, most generally defined, hydrolyze lipids; however, many lipases are active only when the lipid is in micellar form; thus they are characteristically inactive against low concentrations of a lipid, and show activity (surface activation) only when the lipid concentration rises above the critical micellar concentration.

estrogen: (= oestrogen (estrogen))

etioplast: An organelle of a higher plant that has been grown in the dark; it contains no chlorophyll but has chlorophyll precursors and in light can develop into chloroplast.

eu-: A prefix derived from the Greek for good which indicates authenticity.

Eubacteria: (*see* eukaryote)

Eucarya: (*see* eukaryote)

euchromatin: In cytology, the lightly staining regions of chromosomes that contain less-condensed chromatin and are the regions that are transcribed as RNA. (*see also* heterochromatin)

euglobulin: (*see* globulin)

eukaryote: Common name of a member of the *Eucarya*, one of the three domains of living organisms that are classified according to rRNA sequence homologies; they are further characterized by cells containing nuclei and membranes composed primarily of diacylglycerol derivatives. The other two domains are the *Eubacteria* (the bacteria), which have no nuclei (i.e. prokaryotes) and also have diacylglycerol derivative membrane components, and the *Archaea* (archaebacteria), which are also prokaryotes; they have a membrane composed of isoprenyl glycerol diether and tetraether lipids. Kingdoms within the *Eucarya* include plants, animals and fungi; kingdoms of the *Archae* are euryarchaeotes (or euryotes), which include methanogenic, halophilic, sulphur-reducing and some thermophilic organisms, and the crenarchaeotes (or crenotes), which include other sulphur-dependent organisms, thermoacidophiles and extreme thermophiles (eocytes).

euploidy: The state of having the normal number or chromosomes, as contrasted with aneuploidy, having an abnormal number of chromosomes.

euryarchaeote: (*see* eukaryote)

euryote: (*see* eukaryote)

evolutionary tree: (= phylogenetic tree (evolutionary tree))

ex vivo: 'Removed from life', e.g. an explant, a tissue in culture.

EXAFS: (= extended X-ray absorption fine structure (EXAFS))

exaptation: The adaptation over evolutionary time of a protein, and its gene, to a novel function; e.g. the transformation of trypsinogen into an antifreeze glycoprotein.

exchange protein: (= phospholipid-transfer protein (exchange protein))

exchange reaction: A partial enzymic reaction, especially when only one of two substrates is present, in which chemical groups or single atoms of the substrate equilibrate with the medium, with a cofactor or with one of the products, e.g. the acetate-dependent incorporation of $[^{32}P]P_i$ into ATP, catalysed by acetyl-CoA synthetase, in the absence of coenzyme A.

excimer: In fluorescence spectroscopy, a dimer of a fluorophore that is stable only in the excited state, e.g. of aromatics such as pyrene or 2,5-diphenyloxazole (PPO, used for scintillation counting).

excision repair: The removal of a segment of a DNA duplex that contains a thymine or other dimer, formed by ultraviolet-light-induced damage, and resynthesis and ligation of the excised sequence.

excision: The removal of a section of double-stranded DNA that is faulty due to mutation or incorrect replication. The process includes *incision*, i.e. cleavage of the strand by an endonuclease and the action of a 5'-nucleotidase, followed by filling in by a polymerase and ligation. The *E. coli* DNA polymerase I has all these activities. (*see also* proofreading (editing))

excitable membrane: A membrane containing a receptor which transduces an external signal, e.g. a membrane containing acetylcholine receptors or rhodopsin.

excitation spectrum: In fluorescence spectroscopy, emission at a longer, fixed, wavelength as a function of the wavelength of the exciting light. (*see also* emission spectrum)

excitotoxin : A molecular signal that acts via a cell-surface receptor and initiates a series of events that result in cell death; e.g. glutamate, some of whose receptors initiate apoptotic or necrotic cascades.

exergonic: Descriptive of a chemical reaction that generates free energy. (*see also* endergonic; exothermic)

exinuclease: An endonuclease that functions to excise damaged bases, e.g. thymidine dimers.

exo-: A prefix that indicates something external, e.g. an exopeptidase is an enzyme that cleaves N- or C-terminal residues from a polypeptide. (*see also* endo-; ecto-)

exocrine: Descriptive of secretion into ducts that open on to body surfaces rather than into the bloodstream, e.g. salivary and gastric secretions. (*see also* autocrine; endocrine; paracrine)

exocytosis: A secretory mechanism whereby secretory vesicles fuse with the cell membrane and expel their contents to the exterior. (*see also* apocrine; endocytosis; holocrine)

exogenous: Arising from a source outside the organism or cell. (*see also* endogenous)

exon scanning: A procedure for approximating sites of alternate splicing. In the PCR amplification of cDNA, a single primer matched to a sequence near one end, 3' or 5', of the coding sequence, is paired with primers matched to sites increasingly remote from it. The pairs that amplify multiple bands are directed to sites that bracket a site of alternate splicing.

exon shuffling: The result of natural genetic recombination that creates a gene for a novel protein from the exons of existing proteins. The presence in introns of many sites for homologous recombination allows the assembly of a gene for a new protein from the exons of existing genes and introns mark the boundaries between domains. Their products are mosaic proteins; each exon, or sometimes several exons, encodes a separate domain.

exon skipping: A defect in processing of pre-mRNA, due to a mutation that inactivates the splice site between an intron and the exon downstream from it.

exon trapping: A method for recovery of exons from random segments of cloned genomic DNA that depends upon passage through a retroviral life cycle; non-viral genomic DNA is spliced and may then be excised as cDNA from the viral DNA using appropriate exonucleases.

exon: From *expressed region*, i.e. a region of a eukaryotic gene that encodes a sequence of amino acids; as opposed to an *intron*, i.e. an intervening region or sequence. Introns are excised before translation of the resultant RNA transcript.

exonuclease footprinting (DNase footprinting): A method for identification of a protein-binding region in a double-stranded DNA fragment. One 5'-end of the DNA fragment is labelled with ^{32}P and the DNA-fragment-protein complex is treated with a 3'-exonuclease to digest from the 3'-end until it meets the region protected by the DNA-binding protein. The labelled strand is characterized subsequently by its size to indicate the distance of the site of the binding protein from the 5'-end of the fragment. The technique is suitable for localization of a binding site to a specific region of a large DNA fragment, whereas *footprinting* gives the exact sequence of the binding site of a smaller fragment. (*see also* footprinting)

exopeptidase: A peptidase that cleaves a protein sequentially, starting at the N-terminal peptide bond (an *aminopeptidase*) or at the C-terminal peptide bond (a *carboxypeptidase*).

exoskeleton: The chitin-containing rigid external shell of insects and crustaceans.

exosome: A multi-exoribonuclease complex of prokaryotic and eukaryotic cells.

exothermic: Descriptive of a chemical reaction that generates heat. (*see also* endothermic; exergonic)

exotoxin: A disease-causing agent that is produced and secreted by a pathogen; contrasted with an *endotoxin*, which is an intrinsic component of the pathogen, e.g. the lipopolysaccharide of Gram-negative bacteria.

explant: A small sample of living tissue, often consisting of different tissue types, that can be sustained in a relatively differentiated state under cell culture conditions. (*see also* cell culture)

expressed sequence tag (EST): a partial coding sequence isolated at random from a cDNA library; like a sequence-tagged site for mapping genomic DNA, used as a landmark for identification and mapping of coding sequences, for discovery of new genes and, by reference to sequence data banks, for discovery of identities with other genes.

expression cloning: As part of a repetitive process for screening a library, the sampling of cells according to the presence of the product of the gene of interest, e.g. the identification of the glutamate receptor in a clone of cells as evidence that the receptor gene is contained within the cells.

expression vector: A plasmid or bacteriophage, used as a vehicle for transfer of genetic information to a host cell, in which the inserted gene is expressed as a protein, e.g. the gt11 vector.

expression: The production of a gene product, i.e. protein or RNA, from a gene; the manifestation of a genotype as a phenotype.

expression-library immunization (ELI): The use of an expression library constructed from the complete or partial DNA complement of a pathogen to immunize a host without the risk of infection.

extein: (*see* protein splicing)

extended X-ray absorption fine structure (EXAFS): A technique for investigation of the immediate environment of metal atoms in metalloprotein crystals or solutions, e.g. Fe-S bond distances in rubredoxin; the X-ray energy is varied and the fine structure of the absorption spectrum is recorded indirectly as fluorescent radiation.

extinction coefficient: (*see* molar absorption coefficient)

extra-: A prefix derived from the Latin for outside.

extremophile: An organism that can live in an extreme environment, e.g. *barophiles* (or *piezophiles*) that live under high pressure, *psychrophiles* that live at low temperatures, and *thermophiles* and *hyperthermophiles* that live at high temperatures; contrasted with a *mesophile*, which is an organism that lives under moderate conditions.

extrinsic factor: Vitamin B_{12}. (*see also* intrinsic factor)

extrinsic pathway: The blood clotting cascade that is initiated by a tissue factor. (*see also* intrinsic pathway)

extrinsic protein: (= peripheral protein (extrinsic protein))

F

FAB: (= fast-atom bombardment (FAB) mapping)

Fab fragment: A product of papain hydrolysis of an immunoglobulin G consisting of the variable region and some of the constant region of a heavy chain and an entire light chain, and containing a single antigen-binding site. (*see also* Fc fragment)

facilitated diffusion: A transport mechanism that moves compounds or ions down a concentration gradient, and requires no energy; also known as accelerated diffusion or mediated transport. (*see also* active transport; passive diffusion)

FACS: (= fluorescence-activated cell sorting (FACS))

factitious protein: A product of genetic engineering; a protein designed for a specific purpose or for its expected properties.

FARCE: Freezing at the reactive centres of enzymes. (*see* entropy effect)

fast-atom bombardment (FAB) mapping: A technique for comparison of recombinant with natural proteins or with expected structures. A protein is cleaved by a specific enzymic (e.g. trypsin) or chemical (e.g. CNBr) method and the product is characterized by FAB mass spectrometry. (*see also* fast-atom bombardment mass spectrometry (FAB-MS))

fast-atom bombardment mass spectrometry (FAB-MS): A technique for the analysis of protein sequence and structure. A high energy beam of atoms or ions (Cs or Xe) vapourizes, fragments and ionizes a protein solution, the mass spectrum of which gives details of the peptide sequence and post-translational modifications such as N-terminal acylation, glycosylation, phosphorylation and disulphide bridging. *Flow FAB* is a variant that analyses the continuous liquid stream from a high pressure liquid chromatography or capillary electrophoresis separation. (*see also* fast-atom bombardment (FAB) mapping)

fast-twitch muscle: (= white muscle (fast-twitch muscle))

fat: Usually neutral fat; triacylglycerols.

fatty acid: A carboxy group on an alkyl chain that is usually unbranched; may be a short-chain (alkyl group contains less than 7 carbons), medium-chain or long-chain (alkyl group contains more than about 11 carbons) fatty acid.

Fc fragment: A product of papain hydrolysis of an immunoglobulin G consisting of parts of the constant regions of two heavy chains held together by a disulphide bridge, but excluding antigen-binding regions. (*see also* Fab fragment)

FDAT: (= fluorescence-based DNA analysis technology (FDAT))

FD-MS: Field desorption mass spectrometry. (*see* mass spectrometry (MS))

FDR: (= functional divergence ratio (FDR))

feedback inhibition: A form of metabolic control in which the end product of a pathway inhibits the enzyme, usually an allosteric enzyme, that catalyses the earliest irreversible reaction that is unique to the pathway. Also, the control a hormone exerts on the sequence of events that results in its synthesis and release, e.g. the action of thyroxine on secretion of thyrotropin (thryroid-stimulating hormone).

feedback regulation: Control of a metabolic pathway by a metabolite of the pathway that acts in the direction opposite to metabolic flux, i.e. upstream or 'earlier' in the pathway. (*see also* feed-forward regulation)

feed-forward regulation: Control of a metabolic pathway by a metabolite of the pathway that acts in the same direction as the metabolic flux, i.e. downstream or 'later' in the pathway, e.g. the activation of pyruvate kinase by fructose 1,6-bisphosphate. (*see also* feedback regulation)

Fehling's solution: (*see* reducing sugar)

feline: Derived from the cat.

femto-: A prefix (f) which denotes 10^{-15}; e.g. femtomolar, 10^{-15} M.

Fenton reaction: The generation of a reactive, and potentially damaging, hydroxyl radical: $H_2O_2+Fe^{2+}OH \rightarrow +Fe^{3+}OH^-$.

Ferguson plot: A graphical representation of electrophoretic mobility as a function of gel concentration, according to the relationship: $\log M_0/\log M = K_r C$, where M_0 is the free mobility, M is the mobility in a gel of concentration C, and K_r is a retardation constant related to molecular size.

fermentation: The anaerobic degradation, usually by a micro-organism, of a sugar (or other source of energy) and biosynthetic intermediates, during which secondary metabolites may be produced.

ferric chloride test: The chelation of Fe^{3+} by an a-oxo acid to give a coloured product, as a test for metabolic products of amino acids, e.g. with α-oxophenylpyruvate as an indirect test for excess levels of phenylalanine in phenylketonuria.

FFF: (= field-flow (field-force) fractionation (FFF))

FFID: Fission-fragment-induced desorption. (= plasma-desorption mass spectrometry; *see* mass spectrometry (MS))

fibre fluorescence in situ hybridization : A technique that applies *in situ* hybridization methodology to individual DNA fibers in order to map the sequence of genetic markers along them.

fibroblast: A relatively undifferentiated collagen-secreting cell of connective tissue.

fidelity: The property of an amino-acid-activating enzyme or a polymerase to correctly charge a tRNA or to correctly place a residue in a growing polypeptide or polynucleotide.

field desorption mass spectrometry: (*see* mass spectrometry (MS))

field-flow (field-force) fractionation (FFF): A group of techniques for resolution of mixtures of macromolecules or colloidal particles. An external driving force (e.g. gravitational, electrical, magnetic) is applied perpendicular to the direction of continuous flow of the mixture through an elongated chamber, and the deflection of each suspended species in the field directs it to a unique area on the accumulation surface of the chamber.

field-force fractionation: (= field-flow (field-force) fractionation (FFF))

figure-eight form: (*see* Holliday model)

FI-MS: Field ionization mass spectrometry. (*see* mass spectrometry (MS))

fingerprint: The unique pattern of ninhydrin-reactive spots found upon separation of a tryptic digest of a protein by paper electrophoresis in one dimension, followed by chromatography in a second dimension at a 908 angle. Also, any unique pattern that is generated by separation of fragments resulting from limited proteolysis of a protein or hydrolysis of a polynucleotide at specific cleavage sites.

finite decomposition: A type of mathematical modelling of steady-state metabolic rates according to the incremental contributions of all controlling factors. (*see also* sensitivity)

first-order kinetics: In enzymology, the direct dependence of rate on the substrate concentration; seen when this concentration is significantly below the K_m. (*see also* zero-order kinetics)

Fischer convention: A representation of the absolute configuration of an asymmetrical carbon atom, especially the carbons of a sugar, in which the substituents further from the observer are shown above and below and the substituents closer to the observer are shown to the right and left. (*see also* Haworth projection)

FISH: (= fluorescence in situ hybridization (FISH))

flip-flop circuit: The sequence of enzymic events that permits the periodic expression of either of two genes, but never both together, e.g. control of the expression of two of the flagellar proteins of *Salmonella*.

flip-flop (transverse diffusion): The slow diffusion of membrane lipids from one leaflet of a lipid bilayer to the other.

flippase: A putative enzyme that transports membrane elements, e.g. the lipopolysaccharide of Gram-negative bacteria, from the inner to the outer leaflet of the lipid bilayer cell membrane.

florigen: A hypothetical molecule that signals floral induction; i.e. the switch from the vegetative to the reproductive developmental pathway.

flow cytometry: The fluorescence-activated cell sorting technique applied to analysis and/or separation of cells by their physical and biological characteristics.

flow FAB: Flow fast-atom bombardment. (*see* fast-atom bombardment mass spectrometry (FAB-MS))

fluctuating embryo: (*see* framework model)

fluid mosaic model (Singer-Nicolson model): A conceptualization of a plasma membrane as a lipid bilayer in which embedded proteins are mobile in two dimensions. (*see also* Danielli-Davson model; Gortner and Grendel model; unit membrane Danielli-Davson model)

fluor: (= fluorophore (fluor))

fluorescence correlation spectroscopy: The observation of fluctuations in the concentration of sparse fluorophores in a small observation area. Applications include estimation of transport coefficients as a fluorophore repopulates a bleached area, and of very low concentratons as individual molecules diffuse in and out of the experimental field of vision.

fluorescence in situ hybridization (FISH): A method for visualization of a genetic marker on a chromosome by use of a fluorescent labeled polynucleotide probe that hybridizes to and indicates the locus of a gene on a chromosome during mitosis (metaphase). The methodology may also be used to locate and semi-quantitatively estimate the level of a specific mRNA at the sub-cellular level. (*see also* interphase nucleus mapping)

fluorescence lifetime: The average time that a population of fluorophores spends in the excited state before collapse to the ground state. The lifetime is equal to the reciprocal of the first-order rate constant for the decay of fluorescence; this constant is increased, and the lifetime decreased, by non-radiative interaction of the excited state with its environment.

fluorescence photobleaching recovery: A technique to measure the mobility of membrane lipids by destruction of fluorophore-tagged molecules in a very small area of a membrane and measurement of the rate at which new molecules migrate into it from the non-affected margins.

fluorescence recovery after photobleaching: (= fluorescence photobleaching recovery)

fluorescence resonance energy transfer: (= resonance energy transfer)

fluorescence: The property of a compound or moiety of absorbing ultraviolet or visible light and re-emitting it nearly instantaneously at a longer wavelength; also the fluorescent emission itself.

fluorescence-activated cell sorting (FACS): A technique for separation of cells according to their fluorescence after attachment of a fluorophore to specific cells, e.g. with a fluorescent antibody. Droplets that contain no more than one cell are passed by a device that imposes an electrical charge on fluorescent droplets and deflects them into their own receptacle. (*see also* cytoflow; flow cytometry)

fluorescence-based DNA analysis technology (FDAT): Automated DNA sequencing using single-lane sequencing; adaptable to automated data collection.

fluorography: A variant of autoradiography with improved sensitivity for detection of radioactivity of bands in slab gels. The weak emission from ^{3}H, ^{14}C or ^{35}S excites a scintillant such as salicylate that in turn exposes the photographic film.

fluorophore (fluor): A chemical group responsible for the fluorescence of a compound or macromolecule.

flush ended: (= blunt ended)

flux-generating step: An enzymic reaction that limits the rate of subsequent steps in its metabolic pathway. In a steady state the flux of metabolites, expressed as mmol/min, is the same for all steps. The step is usually characterized by thermodynamic irreversibility, and the enzyme has a K_m that is low compared with levels of its substrate *in vivo*, thus insulating itself and the pathway from variations in substrate levels.

foldability: (*see* designability)

foldamer: A polymer of non-natural residues that has peptide-like folding properties; e.g. oligomers of - or -amino acids, which assume stable helical secondary structures in solution.

folding unit: A nucleation site of protein naturation or renaturation.

Folin-Ciocalteu reagent: A reagent that uses phosphomolybdotungstate for the colorimetric detection of proteins. (*see also* Lowry protein assay)

footprinting: A group of techniques for identification of protein-binding sites on DNA. In *dimethylsulfate (DMS) footprinting*, for example, an end-labelled DNA fragment is partially methylated at guanidine residues with DMS, then exposed to the DNA-binding protein. The material is electrophoresed on a non-denaturing polyacrylamide gel to separate fragments that are methylated in the protein-binding sequence (and will no longer bind) from those that are methylated only elsewhere and will be less mobile due to the bound protein. The two gel bands are then eluted and cleaved with piperidine at the methylated sites. Comparison of the cleavage ladders of the two preparations will show unique bands due to methylation and cleavage at residues in the non-binding material that were essential for binding. In *protection footprinting* a DNA fragment with and without a binding protein is subjected to cleavage in non-binding areas, e.g. by partial methylation and cleavage by piperidine. The material is separated on a sequencing gel and the ladder of the protein-bound preparation will show a gap (the footprint) in its sequence where the protein has bound, compared with the ladder from 'naked' DNA. Protection footprinting may be applied to intact cells and compared with naked DNA *in vitro* as the control. In that case, ligand-mediated PCR is used to label and amplify desired areas of the DNA. Other methods of cleavage are by DNase I and by hydroxyl radicals ($OH^{\bullet}$). (*see also* genetic footprinting)

forced copy-choice model: One of two modes of retroviral recombination. In making a DNA transcript of the two genomic

RNA strands of a retrovirus, when the polymerase encounters a break on one template strand it continues on the other strand. More generally, the term describes recombination that occurs during minus strand synthesis. The other mode of retroviral recombination is the *strand displacement-assimilation model*, or recombination during plus strand DNA synthesis: when synthesis occurs on an intact RNA template, one fragment of DNA produced during the semi-discontinuous synthesis is displaced by the DNA strand being continuously synthesized on the other RNA template. Both modes of recombination have been detected.

founder effect: The prominence in a population of a rare allele or haplotype due to the population's being historically isolated and having been descended from a small number individuals, among whom was one individual, the founder, who by chance possessed the rare allele or haplotype.

Fourier transform infrared spectroscopy (FT-IR): The special case of the application to the infrared spectrum (10^{-6}-10^{-4} m) of interference optics. To avoid the disadvantage of dispersion/slit optics (the use of a prism or grating to separate polychromatic light into a spectrum and the exclusion of all the light except that of the desired wavelength), which utilizes only a very small fraction of the incident light, interference optics is used to decompose the absorption of polychromatic light. White light is split into two beams when it strikes a semi-transparent mirror; a system of mirrors rejoins the two beams at a detector. The pathlength of one beam may be adjusted by use of a moveable mirror so that this beam of light constructively interferes with, or re-enforces, the other when the two paths are exactly equal, or differ by an exact multiple of the wavelength; or it destructively interferes with it when the two paths differs by an exact multiple plus half the wavelength. The device, therefore, acts as a very narrow bandpass filter. When, under destructive interference conditions, a sample is introduced into one of the beams, the appearance of light at the detector is proportional to the absorbance of light at the selected wavelength by the sample. The intensity of energy at the detector is monitored as the pathlength is continuously varied over an informative range (800-4000 cm^{-1}). The technique has been adapted to microscopic samples to obtain information on protein secondary structure.

Fourier transformation: An analysis based on Fourier's theorem that any periodic function may be reduced to a series of sine and cosine terms, each represented by an amplitude and a phase. The transformation is the casting of data for a periodic phenomenon into such an alternative form, or the reconstruction of the phenomenon from its mathematical expression, e.g. reconstruction of a molecular model of a compound from an analysis of the X-ray diffraction pattern of its crystal, which is a three-dimensional periodic structure. (*see also* phase problem)

Fourier transform-mass spectrometry (FT-MS): A variant of mass spectrometry in which a molecular ion is stabilized in a cyclotron orbit and detected by the frequency of its oscillation. The consequent long half-life and non-destructive detection account for the high sensitivity of the method. (*see also* mass spectrometry (MS))

four-way junction (4WJ): = Holliday junction

FPERT: A variation on the phenol-emulsion reassociation technique protocol that includes formamide.

FR: Framework region. (*see* complementarity-determining region (CDR))

fragment reaction assay: A measure of the transferase phase of protein synthesis independent of mRNA and many ribosomal factors; i.e. the transfer of an amino acid from its position esterified to the 3'-end of a tRNA (or partial digest of a tRNA) to the N-terminus of the growing peptide chain.

frame-shift suppressor: A mutation that circumvents the effects of a frame-shift, e.g. a mutation in the anticodon region of a tRNA that allows it to recognize a tetranucleotide codon.

frame-shift suppressor: A mutation that circumvents the effects of a frame-shift, e.g. a mutation in the anticodon region of a tRNA that allows it to recognize a tetranucleotide codon.

frame-shift: A mutation that throws out of register the normal reading of triplet codons during translation; usually caused by an insertion (or deletion) of one or two nucleotides into (or from) the gene.

framework region: (*see* complementarity-determining region (CDR))

FRAP: Fluorescence recovery after photobleaching. (= fluorescence photobleaching recovery)

free base: The purine or pyrimidine moiety of a nucleoside, nucleotide or nucleic acid that is not attached to the pentose or pentose phosphate moiety. (*see also* base)

free energy: The form of energy that is capable of doing work at constant pressure and temperature; a concept that bridges the requirement that energy (enthalpy plus potential energy) be conserved and that the randomness (entropy) of a system spontaneously increases; expressed as J/mol (cal/mol). (*see also* second law of thermodynamics)

free radical: A molecular species containing an unpaired electron, e.g. the hydroxyl ($OH^{\bullet}$), superoxide ($O_2^{\bullet -}$) and nitric oxide ($NO^{\bullet}$) radicals. (*see also* lipid peroxidation; oxidative stress)

freeze-etching: (*see* freeze-fracture)

freeze-fracture: A technique to prepare cells for scanning electron microscopy; cleavage of a solidly frozen cell, followed by *freeze-etching*, i.e. sublimation of some of the frozen water of the cell and coating the exposed subcellular structures with a film of platinum or carbon.

freezing at reactive centres of enzymes: (*see* entropy effect)

French paradox: The hypothesis that the lower incidence of heart disease in Mediterranean countries is due to ingestion of constituents of red wine, possibly flavinoids, that act to inhibit oxidation of low-density lipoproteins.

Fresco-Alberts-Doty model: (= hairpin loop (Fresco-Alberts-Doty model))

FRET: Fluorescence resonance energy transfer. (= resonance energy transfer)

frictional coefficient: A measure of the size and asymmetry of a molecule in solution, derived from hydrodynamic measurements, e.g. diffusion, ultracentrifugation, mobility in electrophoresis; equal to RT/ND, qE/v, and $M(1\text{-})/NS$, where R is the gas constant, T is the absolute temperature, N is Avogadro's number, D is the diffusion coefficient, q is the macromolecule's charge, E is the electric field strength, v is the rate of electrophoretic movement, M is the molecular mass, is the partial specific volume, ρ is the solution density and S is the sedimentation coefficient.

FT-MS: (= Fourier transform-mass spectrometry (FT-MS))

Fuelgen reaction: A qualitative colorimetric method for identification of DNA, especially in cytochemistry; treatment with fuchsin sulphurous acid to produce a red colour.

Fugu rubripes: The puffer fish; an attractive experimental animal because it has a genome about one-eighth the size of mammalian genomes, largely due to shorter introns and less repetitive elements.

functional cloning: (*see* positional candidate approach)

functional divergence ratio (FDR): A measure of the evolutionary selective pressure on a protein. In a family of homologous proteins, or the genes encoding them, the amino acid residues are categorized as functional (e.g. those at the binding site and catalytic site) or non-functional. The FDR is the ratio of mutations found among functional residues to those found among non-functional residues.

funnel concept: The new view of the protein folding problem in which the assortment of unfolded polypeptides are considered as an ensemble, the myriad of individual members of which achieve the native form by microscopically unique routes, all funneling down from the high-energy unfolded state to the low energy native conformation. If these molecules, on average, do not encounter an obstacle to refolding, the kinetics appears like the two-state case of the pathway model. Obstacles may be dead ends (kinetic traps) from which the miss-folded protein is extracted by Browninan forces or helped by a chaperonin to reverse to a higher energy state from which it can continue its folding. Another kind of bottleneck is a conformational entropy barrier, when there are only small differences in energy among the various possible conformations each folding polymer can achieve. It would then take some time, on the average, for the partially folded molecules to search the many possible conformations to find a continuation of the folding sequence. By viewing the energy co-ordinate as an energy landscape, folding of specific unfolded proteins may be viewed as passing over plateaus, down steep ravines or trapped in moats. Although the funnel concept, or 'new view' of protein folding was proposed as an alternative to the hypothesis that each protein has only one folding pathway (the 'old view), there in fact may be multiple pathways, but they share many features. *see* framework model; kinetic partitioning; paradox of Levinthal

furanose: The form of a sugar when it is condensed into a five-membered ring. It consists of four carbon atoms and the oxygen atom that is the link to the anomeric carbon atom. (*see also* pyranose)

futile cycle (substrate cycle): The coupling through common substrate/product pairs of an energy-requiring enzymic reaction with another energy-producing reaction that regenerates one of the products of the first reaction. The net result is the expenditure of energy, e.g. the phosphofructokinase and fructose phosphate phosphatase reactions that together generate and hydrolyse fructose bisphosphate with the net consumption of ATP. Such a cycle is potentially useful in the rapid response of metabolism to regulatory factors.

G

-gen: -gen, -genic, -genesis. Suffixes with the same derivation as gen- and which indicates a precursor, e.g. pathogen, 17-ketogenic, glucogenesis.

G band: (*see* AT queue)

G loop: (*see* AT queue)

G1 period: (*see* cell cycle)

G2 period: (*see* cell cycle)

G-4 DNA: *see* quadruplex DNA

GABA shunt: An alternative route from α-oxoglutarate to succinyl-CoA as part of the tricarboxylic acid cycle. Rather than going through the α-oxoglutarate dehydrogenase reaction, α-oxoglutarate condenses with NH_3 to form α-iminoglutarate, then is reduced to glutamic acid, decarboxylated to γ-aminobutyrate (GABA), transaminated to succinic semialdehyde and oxidized to succinyl-CoA.

gain-of-function: *see* loss-of-function

gamete imprinting: (*see* imprinting)

gamete: A spermatozoon or ovum; a haploid germ cell.

gamma (γ)-globulin: (*see* globulin)

gap junction (cell-to-cell channel): A connection between adjacent cells that allows small molecules and ions, e.g. amino acids, sugars and nucleotides, to pass from the cytoplasm of one cell to the cytoplasm of the other.

gas-phase analogy: (*see* entropy effect)

gas-phase sequencing: (*see* solid-phase sequencing)

gastrulation: During early embryogenesis, the invagination and reshaping of the cells of the blastoderm that results in differentiation into ectoderm, endoderm and mesoderm.

gate: A regulated channel through which ions pass across the plasma membrane; *gating* is the opening and closing of the channel.

gatekeeper protein: A protein that monitors transfer of a protein from the endoplasmic reticulum to the Golgi apparatus and prevents transfer of newly synthesized proteins with inappropriate conformations or with unpaired thiol groups.

gating: (*see* gate)

GATT: (= gene amplification with transcription/translation (GATT))

gauche conformation: An arrangement in space of the carbon-carbon bonds of the alkyl chains of a fatty acid residue of a membrane lipid. The backbone of the chain is rotated 120° from the *trans* conformation, to force the backbone into making a 60° turn through the bond.

GAWTS: (= genomic amplification with transcript sequencing (GAWTS))

GC clamp: (*see* denaturing gradient gel electrophoresis (DGGE technique))

gd-PCR: (= gene dosage PCR (gd-PCR))

gel filtration chromatography: A technique for separation of soluble compounds according to size, based on their relative ability to penetrate the carefully sized pores of the stationary gel phase.

gel retardation assay: (*see* gel shift assay (electrophoretic mobility shift assay; EMSA))

gel shift assay (electrophoretic mobility shift assay; EMSA): A method to identify by function a band on a polyacrylamide gel. Electrophoresis is performed on a mixture of proteins before and after treatment with a strong ligand that is specific to one component, and note is made of the absence or shift in position of a protein due to prior binding to the ligand. A variant is the *gel* *retardation assay*, an approach to detecting DNA-binding proteins by their ability to bind to a short radiolabelled DNA fragment and, as a result, decrease its electrophoretic mobility in a gel.

gen-: A prefix derived from the Greek for born and which indicates a connection to the propagation of life or to heredity.

gene amplification with transcription/ translation (GATT): A method for the production of up to 10^{10} protein molecules per DNA by a continuous process in a homogeneous solution.

gene amplification: The synthesis of multiple copies of a specific gene, which remain as tandem repeats within the chromosome or are segregated as satellite DNA. Amplification is usually not inherited, in contrast to gene duplication.

gene conversion: The replacement of an allele during crossing- over owing to mismatch repair of the heteroduplex intermediate; rather than being equally represented in daughter DNA strands, one allele is over-represented and the other is under-represented. Partial repair can result in a hybrid or mosaic gene.

gene dosage PCR (gd-PCR): A method for quantification of the relative amount of a gene in a biological sample. One primer is constructed to hybridize with a gene known to be in the chromosome in question (e.g. chromosome 21 in the case of Down's syndrome) and another to hybridize to an unrelated gene in a different chromosome. Quantification of the amplified PCR products is sufficiently precise to distinguish a 50% increase in the chromosome 21 gene.

gene duplication: The inherited appearance of multiple copies of a gene. Duplication may result in identical or defective copies, either as tandem repeats at the original chromosomal locus, at a neighboring locus or at a completely different locus. Duplication may be benign or may be the basis of a pathology. If benign, one of the copies will be freed of any selective pressure and may evolve a new function. *see* gene amplification

gene probe: (= hybridization probe)

gene product: The result of gene expression, i.e. an RNA or protein molecule.

gene targeting: The introduction of a homologous DNA sequence into a specific site in the genome of a cell, either by replacement of the former sequence, i.e. *sequence replacement*, or by insertion into the former sequence, i.e. *sequence insertion*. A vector that is homologous and co-linear with a partial sequence of the targeted gene is introduced into an appropriate cell, e.g. a stem cell of some sort; it is incorporated into the genome of some of the cell's progeny and replaces the former sequence. If, however, homologous regions of the vector hybridize to the gene, and it is then incorporated into the gene by homologous recombination, the result is a disruption of the gene by insertion of the vector sequence.

gene therapy: The *in vivo* introduction of a functional gene, and its expression, in the germ line cells (*germ line therapy*) or somatic cells (*somatic gene therapy*) of an individual who does not possess the normal gene.

gene: In genetics, a unit inferred from the pattern of inheritance; in molecular biology, defined narrowly as a section of DNA that is expressed as RNA or, more widely, as a coding sequence of DNA and associated regulatory sequences.

general acid: In chemistry, a proton donor that can participate in catalysis. (*see also* general base; specific acid)

general base: In chemistry, a proton acceptor that can participate in catalysis. (*see also* general base; specific acid)

general genetic recombination: Recombination between homologous chromosomes that may occur during meiosis anywhere along their lengths. (*see also* Holliday model; Meselson-Radding model)

genetic code blocker: (= code blocker (antisense drug))

genetic code: The series of codons of mRNA, each of which specifies a single specific amino acid.

genetic dissection: An approach to defining the roles of individual factors in a complex system. Mutations in a metabolic pathway or physiological function are selected and the function of the cognate gene product is analysed. First addressed to metabolism in fungi, genetic dissection has more recently been applied to phenomena such as phototransduction in *Drosophila* retinas. *see* one-gene one-enzyme hypothesis

genetic fingerprinting: A technique for establishing genetic relationships, especially in forensic medicine, by comparison of the occurrence of uncommon genetic markers.

genetic footprinting: An approach to the identification of the function of a large number of putative genes of a micro-organism, such as may have been uncovered by genome sequencing. A transposable element is inserted into a large number of sites in the genome of the micro-organism, and the mutagenized population is then grown under a wide variety of conditions that may suggest a gene's function. After many population doublings under each set of conditions, the micro-organisms' DNA is extracted and used as a resource for analysis of as many of the putative genes as is desired. PCR primers are constructed to hybridize with the transposable element and with a putative gene. PCR amplification using the primer pair will produce a spectrum of bands on separation by polyacrylamide-gel electrophoresis, one from each mutation that can grow under the set of conditions, i.e. a genetic footprint of the gene. A comparison of the gene's footprints under permissive and restrictive conditions will indicate, by the absence of bands in the latter footprint, those growth conditions that require the function of the gene.

genetic map: (*see* mapping)

genetic marker: A DNA sequence that can be recognized and thus used to characterize the larger DNA sequence and the chromosome in which it occurs.

genome: The genetic complement of an organism represented by its DNA or, in some viruses, its RNA.

genomic: Descriptive of the DNA of an organism. Genomic DNA, which contains introns, is contrasted with cDNA, complementary DNA produced by reverse transcription of mRNA. (*see* genomics)

genomic amplification with transcript sequencing (GAWTS): A DNA sequencing method that involves PCR amplification of the target and uses a primer with an attached phage promoter sequence, transcription to produce a single-stranded RNA and reverse transcription with dideoxy terminator nucleotides that upon electrophoresis will generate a sequencing ladder.

genomic DNA: DNA that has been isolated from a cell and therefore contains introns, as opposed to cDNA.

genomic library: A collection of transformed cells, each of which contains DNA fragments; the entire population represents the total genome of an organism, e.g. a rat library containing DNA fragments which together comprise the entire rat genome. Appropriate screening methods can select a single transformed cell that contains a specific gene. (*see also* cDNA library)

genomic mismatch scanning (GMS): A method to detect and isolate DNA sequences that are candidate genes for inherited disorders for which the gene product is unknown, based on the absence of mismatches in DNA sequences between an affected individual and a heterozygous or carrier progenitor. Large DNA segments are prepared from the genomic DNA of the two related individuals in such a way (e.g. by leaving 3'-overhangs) that they will not be degraded by subsequent exonuclease digestion, e.g. by *Exo*III. The DNA from one individual is enzymically methylated and annealed with the DNA of the second, heterohybrids (two methylated or two unmethylated strands) are cleaved by appropriate endonucleases, e.g. *Dpn*I and *Mbo*I, and the uncleaved duplexes are scanned for single-base mismatches by methyl-directed mismatch repair enzymes that leave single-strand nicks that are attacked by *Exo*III. The duplexes that survive all these tests are those that are shared by the two related individuals and are therefore candidates for the affected gene.

genomic SELEX: A variant of the SELEX (systematic enrichment of ligands by experimental enrichment) procedure, that uses protein ligands and genomic DNA fragments from the same organism to select RNA-protein binding pairs. (*see* cyclic amplification and selection of targets)

genomic subtraction: (= subtractive DNA cloning)

genomic tag model: The hypothesis that a tRNA-like loop (the tag) of the genomic RNA of some viruses that functions in copying a strand of RNA into a complementary RNA or DNA strand is a relic of the part of the RNA that functioned in initiating its own replication in the primordial RNA world.

genotoxin: A mutagen.

genotype: The genetic composition of a cell or organism, usually with reference to a particular set of alleles that may be homozygous or heterozygous for a trait; the *phenotype* is the appearance of the cell or individual due to actual expression of the alleles that are present.

ghost: An erythrocyte plasma membrane; the product of red cell lysis.

gibberellin: A type of plant hormone; a diterpene that is responsible for cell differentiation, e.g. gibberellic acid.

Gibbs-Donnan effect: (= Donnan effect (Gibbs-Donnan effect))

giga- (G): A prefix which denotes 10^9; e.g. gigabase pair, 1 Gbp, 10^9 bp

global regulator: A metabolite that regulates many operons, e.g. glucose, as it regulates several anabolic reactions in a bacterium.

global regulatory circuit: A series of operons that are induced together, e.g. the genes for the heat-shock proteins or those for the SOS repair proteins.

globular: Descriptive of the folding of a protein upon itself in several convolutions to create interactions of the side chains in many weak salt bridges, hydrogen bonds and hydrophobic bonds that result in a roughly spherical (globular) molecular shape. Single-stranded RNA may also accept a globular shape, and shares with proteins the property of selectivity in binding of ligands.

globulin: Archaic nomenclature for a protein that is sparingly soluble in water, but is soluble in dilute salt solutions. *Euglobulins* do not dissolve in salt-free water, whereas *pseudoglobulins* are soluble in salt-free water. *Gamma ()-globulins* are a population of globulins defined further by their electrophoretic behaviour, and include the immunoglobulins. (*see also* albumin)

glucocorticoid: An adrenal steroid that increases blood glucose concentration, e.g. cortisol.

glucogenic amino acid: An amino acid whose carbon skeleton can be metabolically converted, at least in part, into glucose. (*see also* ketogenic amino acid)

gluconeogenesis: The biosynthesis of glucose from smaller, non-carbohydrate, metabolites, i.e. amino acids, tricarboxylic acid cycle intermediates, lactate or glycerol.

glucose effect: The repression of some bacterial enzymes by the presence of glucose in the growth medium. Depression of cyclic AMP levels due to the availability of glucose is mediated by inhibition of adenylate cyclase by a glucose catabolite. In the absence of glucose, a complex between cyclic AMP and its binding protein attaches to specific regions of DNA and activates transcription. Cyclic AMP acts as a starvation signal, as it does in animal cells, but in bacteria it acts at the transcriptional level. (*see also* diauxie; positive control)

glucose repression: (*see* glucose effect)

glycation: The reversible condensation of the carbonyl group of a sugar with an amino group of a protein to form a Schiff base that often triggers subsequent irreversible rearrangements, e.g. the condensation of glucose with haemoglobin A to form haemoglobin A_{1c}.

glycation: The reversible condensation of the carbonyl group of a sugar with an amino group of a protein to form a Schiff base that often triggers subsequent irreversible rearrangements, e.g. the condensation of glucose with haemoglobin A to form haemoglobin A_{1c}.

glyco-: A prefix derived from the Greek for sweet which refers primarily to sugars.

glycobiology: The study, at the molecular level, of the structural and functional roles of carbohydrate-containing structures and of the interactions of carbohydrates with other structures, especially proteins. (*see also* glycoconjugate)

glycocalyx: The carbohydrate coating on the external surface of a cell membrane.

glycoconjugate: A compound composed of an oligosaccharide linked to a protein or lipid, e.g. mucins (which are glycoproteins), globosides (which are *N*-acetylsphingosine conjugates of oligosaccharides), and the endotoxin of Gram-negative bacteria. (*see also* lipopolysaccharide (LPS))

glycoform: One of several variants of a glycoprotein that have identical polypeptide moieties but differ in the site of glycosylation (site heterogeneity) and/or the nature of the oligosaccharide moiety. Glycoforms are characteristic of the tissue of origin, or of the disease or mutagenic state.

glycogen shunt: A proposed metabolic pathway for the synthesis, in the brain, of ATP from circulating glucose. Rather than being phosphorylated by hexokinase in preparation to be degraded via the glycolytic pathway, blood glucose is incorporated into glycogen and upon metabolic demand, enters glycolysis via the phosphorylase and phosphoglucoisomerase reactions. Although this is less energetically efficient, it can generate ATP faster than the hexokinase route.

glycogen storage disease: A defect in the turnover of glycogen that leads to increased amounts of the normal or abnormal polysaccharide in affected cells.

glycolipid: A compound with both lipid and sugar moieties, e.g. lipopolysaccharide of Gram-negative bacteria cell walls, blood group antigens, gangliosides (carbohydrate moieties attached to N-fatty acylsphingosine), cerebrosides and ceramides.

glycolysis: One of the central pathways of metabolism in most eukaryotes and many other cells; the sequence of enzymic reactions that converts glucose into lactic acid (*anaerobic glycolysis*) or into pyruvate (*aerobic glycolysis*).

glycoprotein hormone: A representative of a class of hormones that share nearly identical A-chains and unique B-chains, each with about 125 amino acid residues and N-linked or O-linked carbohydrate moieties, e.g. thyrotropin, follitropin.

glycoprotein: A protein with one or more carbohydrate moieties attached to it.

glycosaminoglycan: A subunit of a proteoglycan; an oligosaccharide that contains repeating disaccharide residues. One half of each disaccharide is an amino-sugar residue. Glycosaminoglycans include the sulphated chondroitins, dermatans and keratans, and the unsulphated hyaluronans. (*see also* ambidexteran; core protein; link protein; proteodermochondran sulphate)

glycoside: A metabolite formed by conjugation with a sugar through its anomeric carbon atom, e.g. β-methyl glucoside. (*see also* aglycone)

glycosidic bond: The linkage of a sugar hemiacetal or hemiketal through its anomeric carbon with another moiety. (*see also* glycoside)

glycosylation: Post-transcriptional modification of a protein by the addition of a carbohydrate moiety. (*see also* N-linked carbohydrate; O-linked carbohydrate)

glyoxylate cycle (Krebs-Kornberg cycle): A series of metabolic reactions in plants and bacteria, the net result of which is production of a 4-carbon compound from two 2-carbon compounds. The unique reactions are catalysed by malate synthase and isocitrate lyase.

glyoxysome: An organelle in plant cells in which the glyoxylate cycle operates.

glypiated: Descriptive of a glycosylphosphatidylinositol (GPI)-anchored protein, i.e. one that is associated with a cell membrane through an attached GPI group. An alternative term is *piglylated*, from phosphatidylinositol glycan (PIG).

glypican: (*see* proteoglycan (mucopolysaccharide))

GMS: (= genomic mismatch scanning (GMS))

Golgi apparatus: A stack of flattened vesicles that functions in the post-translational processing and sorting of proteins. The Golgi receives proteins from the rough endoplasmic reticulum (RER) and directs them to secretory vesicles, lysosomes or the cell's plasma membrane. The movement of the proteins takes place by transfer vesicles that bud off from the RER or Golgi and fuse with the Golgi, lysosomes or plasma membrane. The surface of the Golgi that faces the RER is termed the *cis*-Golgi, the surface towards the apical membrane is the *trans*-Golgi, and the intermediary part is the medial Golgi.

Gortner and Grendel model: A model in which erythrocyte membranes comprise only phosphoacylglycerols, i.e. no protein or sterol component is envisaged; the model presents a bilayer in which each leaflet is composed of lipids stacked perpendicular to the plane of the membrane, with their alkyl chains oriented inwards (towards the opposing leaflet) and with the polar groups oriented towards the exposed surfaces. (*see also* Danielli-Davson model; fluid mosaic model (Singer-Nicolson model); unit membrane Danielli-Davson model)

gp: A prefix that indicates a glycoprotein; e.g. gp85, a glycoprotein with the mobility in SDS gel electrophoresis of an 85-KDa molecule.

GPI: Glycosylphosphatidylinositol. (*see* glypiated)

G-protein: GTP-binding protein; one of a superfamily of proteins that function in, for example, signal transduction (e.g. the G-protein associated with the b-adrenergic receptor), polymerization (e.g. tubulin), ribosomal protein synthesis (e.g. the translocase), cell differentiation (ras proteins) and intracellular transport of proteins, vesicles or cytoskeletal elements (e.g. dynamin). The common functional feature of the family is an affinity for a target protein in the presence of GTP which is lost upon hydrolysis to GDP.

G-quadruplex: quadruplex DNA

G-quartet: (*see* quadruplex DNA)

Gram-negative: Descriptive of bacteria that do not stain by the Gram method, i.e. bacteria with two membranes. (*see also* Gram-positive; lipopolysaccharide (LPS))

Gram-positive: Descriptive of bacteria that stain by the Gram method, i.e. bacteria with single membranes. (*see also* Gram-negative)

granulocyte: A granular leucocyte; a basophil, eosinophil or polymorphonuclear leucocyte (neutrophil).

granum: A closely packed stack of thylakoid membranes in a chloroplast.

granzyme: A T-cell (granulocyte) proteinase.

-graph: A suffix derived from the Greek for written.

graphi-: A prefix with the same derivation as -graph

Greek key: A topology found in some protein structures; antiparallel β-strands are organized into a barrel in which strands that are adjacent in the primary structure are not always adjacent in super-secondary structure but, when represented in two dimensions, form a Greek key pattern.

green beard: A conspicuous phenotype that allows behavioural differentiation among individuals of the species. When the consequent behaviour is beneficial, the encoding gene can be considered a selfish gene.

gRNA: (= guide RNA (gRNA))

ground state: The unexcited electronic state. (*see also* singlet state; triplet state)

ground substance: The histologically featureless extracellular matrix of connective tissue, largely composed of proteoglycans.

group I self-splicing: (*see* triplet state)

group II self-splicing: (*see* self-splicing)

group transfer hydration: (*see* entropy effect)

group transfer potential: A quantitative measure of the strength of attachment of a moiety (the group) to the rest of the molecule; the difference in standard free energies between the group attached to the molecule and the group attached, usually, to water, e.g. the standard free energy (ΔG^{o}') of the hydrolysis of ATP to ADP and phosphate, -30.5kJ/mol (-7.3kcal/mol).

growth cone: In a developing or regenerating neuron, the leading edge of a focus of dendrite extension, which is responsive to environmental cues that determine its growth rate and the direction of that growth.

growth factor: A protein that binds to receptors on specific cells and promotes their growth, e.g. nerve growth factor, epidermal growth factor, platelet-derived growth factor.

G-tetrad: (*see* quadruplex DNA)

G-tetraplex: (*see* quadruplex DNA)

GTPase protein: (= G-protein)

GTP-binding protein: (= G-protein)

guide RNA (gRNA): A transcript of maxicircle or minicircle DNA, that in lower eukaryotes is thought to function in RNA editing, i.e. by the addition or deletion of U residues at specific sites in the coding regions of mRNA, itself derived from maxicircle DNA as an untranslatable cryptogene, the creation of a translatable open reading frame. The gRNA attaches to an unedited region (pre-edited region, PER) which at a GA-rich sequence anchors the oligo-U 3'-end of the gRNA and at a downstream sequence anchors the complementary 5'-end of the gRNA. Between the two anchored regions of the gRNA, a non-complementary oligonucleotide sequence loops out. An endonuclease cleaves the PER between the two anchors and uses a G residue of the gRNA loop as a template to patch in a U residue. The gRNA is then free to migrate to another site for editing. In some cases this mechanism inserts nearly half the U residues necessary for translation of a mRNA. This extensive editing of a single mRNA is termed pan-edition.

guide sequence: In the self-splicing of RNA, the purine-rich region of the intron that binds a pyrimidine-rich region of the upstream exon to the site where transesterification takes place.

gynogenic: (*see* anthrogenic)

H^+: ATP ratio: (*see* P:O ratio)

H+: O ratio: (*see* P:O ratio)

haemato-: A prefix derived from the Greek for blood.

haemo-: A shortened form of haemato-

haemoglobinopathy (hemoglobinopathy): An abnormality that may produce a disease the basis of which is an alteration in haemoglobin structure or synthesis.

haemolysis (hemolysis): The bursting of erythrocytes due, for example, to a hypotonic environment.

haemostasis (hemostasis): The stopping of blood loss from the vascular system by vasoconstriction, formation of a platelet plug and blood clotting.

hairpin bend: (= reverse turn)

hairpin loop (Fresco-Alberts-Doty model): A region in a single polynucleotide strand containing adjacent complementary sequences that allows the polynucleotide to fold back upon itself and form a double-stranded section, as in the cloverleaf structure of a tRNA molecule.

hairpin RNA: (*see* ribozyme)

half site: One of the two symmetrical duplex moieties of a palindromic site in DNA.

half-chair conformation: The conformation of a pyranose ring that has been strained to place four adjacent atoms in one plane, e.g. the postulated transition state of the *N*-acetylglucosamine residue during lysozyme catalysis, in which C-1 becomes a planar carbocation.

half-life: When a property (e.g. radioactivity, enzyme activity, etc.) decreases at a rate proportional to its concentration, the time, $t_{1/2}$, taken for the property to decrease to one-half its initial value; related to the first-order rate constant, *k*, by $t_{1/2}=\ln 2/k$.

half-of-the-sites activity: An extreme form of negative co-operativity shown by a multisubunit enzyme in which, as a substrate binds to the active site of one subunit, the affinity for the substrate of an adjacent subunit decreases.

halophilic: Descriptive of organisms that survive and grow only in high salt concentrations, e.g. >3 M NaCl. (*see* osmotolerant)

halt: The cessation of transcription due to the absence of a nucleoside triphosphate substrate (in describing the action of RNA polymerase). This is contrasted with arrest, the cessation of transcription due to the absence, or presence, of a control factor, possibly when transcription arrives at a specific site on the template.

hammerhead RNA: (*see* ribozyme)

handshake model: (*see* induced-fit theory (rack model))

handshake: An interaction of two identical or nearly identical structures, proteins or nucleic acids, which fit each other in a symmetrical fashion.

hanging-drop technique: A method for slowly changing concentrations in a protein solution in an attempt to prepare crystals. A small volume of protein solution (of the order of 5 l) on a microscope coverslip is sealed drop-side down over a small well containing a larger volume (of the order of 1 ml) so that, over time, volatile solvents will exchange between the hanging drop and the solution in the well via the vapour phase and concentrate the protein solution or allow alcohol to pass over to it. The progress of crystallization is observable in the droplet through the coverslip.

haploid: Descriptive of the number of chromosomes of a gamete; one of each pair of chromatids. (*see also* diploid)

haplotype: A characteristic combination of alleles on a single chromosome, which may persist in a population because of their close proximity on the chromosome, a founder effect or natural selection.

hapten: A small molecule which by itself is not an antigen, but which as a moiety of a larger structure (a *haptenic determinant*) can serve as an antigenic determinant.

haptenic determinant: (*see* hapten)

Harden-Young ester: Obsolete name for fructose 1,6-bisphosphate

HAT medium: (= hypoxanthine/aminopterin/thymidine (HAT) medium)

Hatch-Slack pathway: (= C_4 photosynthesis (C_4 cycle; Hatch-Slack pathway))

Haworth projection: A two-dimensional representation of pyranose and furanose structures in which they are shown as hexagons and pentagons respectively. The lower ring atoms are understood as being towards and the upper ones away from the observer; the substituents of ring carbon atoms are shown directly above and below the apices of the polygon. (*see also* Fischer convention)

HCA: (= hydrophobic cluster analysis (HCA))

H-DNA: A form of triplex DNA characterized by Hoogstein base pairing.

headward: Descriptive of a mechanism of polymerization. The head of a biosynthetic monomer is defined as the chemically activated site (i.e. the carbonyl groups of amino acyl-tRNA or fatty acyl-CoA esters, the hemiacetal oxygen of a UDP-monosaccharide, the a-phosphate of a nucleoside triphosphate, the alkyl oxygen of isopentenyl pyrophosphate), and the tail as the other site on a monomer that will be joined to an adjacent moiety. Headward elongation occurs when a monomer condenses by displacement of the activating group (tRNA, CoA, pyrophosphate) of the growing polymer (protein, fatty acid or terpene biosynthesis). Conversely, *tailward* elongation occurs when the condensation removes the activating group of the newly added monomer (nucleic acid or polysaccharide biosynthesis).

heat-shock consensus element (HSE): A 14 base pair polynucleotide sequence upstream from structural genes for heat-shock proteins, required for optimal expression. A somewhat longer polynucleotide sequence that includes flanking regions of the HSE is the *heat-shock regulatory element.*

heat-shock protein: One of the proteins produced by some cells when they are stressed, e.g. by an abrupt increase in temperature. *see also* **chaperone machine**

heat-shock regulatory element: (*see* heat-shock consensus element (HSE))

heavy chain: One of two or more kinds of polypeptide chain of a heteromultimeric protein, distinguished by molecular mass, e.g. one of the larger polypeptide chains that, when combined with another heavy chain and two light chains, make up an immunoglobulin molecule; also the largest of the polypeptide chains of any multimeric protein with non-identical subunits.

heavy subunit: The larger of the two ribonucleoprotein complexes that make up a ribosome; more generally, the largest of the subunits of any complex. (*see also* light subunit)

heavy-chain library: (*see* repertoire cloning)

heavy-ion-induced desorption: (*see* mass spectrometry (MS))

Heinz body: A precipitate in the cytoplasm of an erythrocyte of a spontaneously oxidized, unstable, haemoglobin. (*see also* inclusion body)

helical wheel: A representation of the array of amino acid residue side chains around an a-helix viewed end-on, down the axis of the helix. (*see also* Eisenberg plot; hydrophobic moment)

helix cap: A sequence of amino acids in a protein that terminates a stretch of -helical structure.

helix-coil transition: In nucleic acid and protein chemistry, the melting, or co-operative thermal breakdown of the hydrogen bonds that stabilize the secondary structure, of the macromolecule.

helix-loop-helix (HLH): A motif found in some DNA-binding proteins.

helix-turn-helix (HTH) motif: A motif found in many sequence-specific DNA-binding proteins, e.g. in the lambda repressor and related proteins. (*see also* HLH protein)

helper T-cell: (*see* T-cell)

hemiacetal: The product of reversible condensation of an aldehyde and an alcohol in which the alcoholic hydroxy group adds across the carbonyl group of the aldehyde, e.g. glucose acting as both an aldehyde and an alcohol to give the internally condensed glucopyranose; also the bond formed by this condensation. An *acetal* is the product formed by abstraction of the hydroxy group of a hemiacetal and the hydrogen of a second alcohol, e.g. a glucopyranoside or a polysaccharide. (*see also* hemiketal)

hemicatenane: (*see* plectonemic)

hemi-channel (connexon): Half of a gap junction. Being embedded in the membrane of each of the cells and composed of six molecules of the integral protein connexin, the hemi-channel allows communication between the cytoplasm of adjacent cells.

hemidesmosome: A structure of the plasma membrane of an epithelial cell that extends into both the cytoplasm and the extracellular space, and anchors the cell to the extracellular matrix of the basal lamina.

hemiketal: The product of reversible condensation of a ketone and an alcohol in which the alcoholic hydroxy group adds across the carbonyl group of the ketone, e.g. fructose acting as both a ketone and an alcohol to give the internally condensed fructofuranose; also the bond formed by this condensation. A *ketal* is the product formed by abstraction of the hydroxy group of a hemiketal and the hydrogen of a second alcohol, e.g. a fructofuranoside. (*see also* hemiacetal)

hemisubstrate: An enzyme inhibitor that acts in the initial phases of enzyme action as a substrate, but does not complete the catalytic cycle to regenerate an active enzyme; e.g. organophosphate ester anti-cholinesterases, which mimic acetylcholine in that the phosphoric acid moiety of the hemisubstrate acts like the acetly group of acetylcholine to form an ester with the enzyme's active-site serine residue but, unlike the acetyl-enzyme intermediate of the normal enzymic action, the phosphoryl-enzyme is only very slowly hydrolysed to regenerate an active enzyme.

hemizygous: Descriptive of a trait that is carried on the X chromosome and is therefore present in the male population at half the frequency that occurs in females.

hemoglobinopathy: (= haemoglobinopathy (hemoglobinopathy))

hemolysis: (= haemolysis (hemolysis))

hemostasis: (= haemostasis (hemostasis))

Henderson-Hasselbalch equation: A logarithmic form of the equation for the dissociation of a weak acid: pH= pK_a+log[salt form]/[acid form].

hepato-: A prefix that indicates the liver, e.g. hepatocyte.

hetairokaryote: The endosymbiotic product of a eukaryotic cell dwelling within another eukaryotic cell; as contrasted with the mitochondrion and chloroplast, which are believed to be endosymbiotic products of a prokaryote within a eukaryote.

hetero-: (*see* homo-)

heterochromatin: In cytology, the heavily staining regions of chromosomes that contain condensed chromatin. (*see also* euchromatin)

heteroduplex analysis (mismatch analysis): A method for detection of single base substitutions in DNA fragments. The PCR-amplified putatively mutated DNA fragment is mixed with homologous normal DNA fragments; any mutated sequence is much more dilute than the normal sequence and will therefore be less likely to re-anneal. They will appear on polyacrylamide-gel electrophoresis as band mobility shifts, especially under mildly denaturing conditions, and will indicate the presence of heteroduplexes.

heteroduplex mapping: An electron microscopic technique that compares sequence relationships of two polynucleotides, e.g. a spliced eukaryotic mRNA with a coding strand of DNA. Introns of the DNA strand appear as D-loops that deviate from the base-paired sections where the two strands are complementary.

heteroduplex: A partially double-stranded polynucleotide in which one strand contains sequences not fully complementary to the opposite strand, e.g. mRNA, which contains no introns, annealed to the coding strand of its corresponding gene, which contains introns. Heteroduplexes are products of recombination between DNA duplexes at a region of heterology. (*see also* D-loop)

heterogeneous: Descriptive of a diversity according to some criterion for detecting it; e.g. a mixture of proteins or polynucleotides of differing electrophoretic mobility, size or sequence. Heterogeneity is contrasted with homogeneity, a uniformity according to some criterion. (*see also* **heterogenous**)

heterokaryon: A hybrid of cells from different species, e.g. a fusion of a human and a mouse cell.

heterolactate fermentation: (*see* homolactate fermentation)

heterolytic cleavage: The splitting of a covalent bond that leaves both bonding electrons with one of the atoms. (*see also* homolytic cleavage)

heteromer: *see* homomer

heteromeric: (*see* homomeric)

heterophagy: The action of a lysosome of a phagocytic cell to digest extracellular materials. (*see also* autophagy)

heterophagy: The action of a lysosome of a phagocytic cell to digest extracellular materials. (*see also* autophagy)

heteropolysaccharide: A carbohydrate composed of more than one kind of sugar monomer.

heteropolysaccharide: A carbohydrate composed of more than one kind of sugar monomer.

heterotrophic: In microbiology, descriptive of organisms that require organic material for growth. (*see also* autotrophic; mixotrophic)

heterotropic enzyme: An enzyme that is controlled by an allosteric activator that is not also a substrate. (*see also* homotropic enzyme)

heterozygote: (*see* allele)

hexose monophosphate shunt: (= pentose phosphate pathway)

hierarchical condensation: The hypothesis that, during the folding of a protein in solution, some regions are much more likely to adopt a secondary structure and that, once formed, these interact to direct the subsequent course of folding. This hypothesis is a component of the framework model.

high-energy bond: A chemical bond whose hydrolysis results in the generation of 30kJ (7kcal) of energy or, if coupled to an energetically unfavourable reaction, can drive that reaction forward.

high-energy phosphate: A phosphate anhydride, enol phosphate or similar compound whose hydrolysis is associated with a large decrease in free energy. (*see also* high-energy bond; phosphagen)

high-mannose-type carbohydrate: One type of glycoprotein moiety that is attached to the -amide nitrogen of an asparagine residue (N-linked), and contains exclusively mannose residues at the non-reducing ends. (*see also* complex-type carbohydrate; hybrid-type carbohydrate)

high-mobility-group (HMG) protein: One of a class of non-histone nuclear nucleoproteins, i.e. acidic proteins of chromatin that are characterized by their elution properties from an ion-exchange column.

high-performance liquid chromatography: (= high-pressure (high-performance) liquid chromatography (HPLC))

high-pressure (high-performance) liquid chromatography (HPLC): A technique for rapid separation of solutes on a solid support, based on ion-exchange, gel permeation or partition principles; *reverse-phase HPLC* is when the stationary phase has a non-polar coat; *microbore* or *narrow-bore HPLC* is adapted to small quantities and high sensitivity by scaling down.

high-voltage electrical discharge: (= electropermeabilization)

HIID: Heavy-ion-induced desorption. (*see* mass spectrometry (MS))

Hill coefficient: A measure of co-operativity, i.e. the number h in the equation $[HbO_2]/[Hb]=k(pO_2)^h$ that empirically describes the dependence of haemoglobin (Hb) oxygenation on the partial pressure of oxygen (pO_2).

Hill plot: A graphical representation of binding data, especially for oxygen binding to haemoglobin; a plot of $\log[Y/(1\text{-}Y)]$ against

$\log pO_2$, where Y is the fraction of binding sites occupied and pO_2 is the partial pressure of oxygen; the slope of the plot is the Hill coefficient (h).

Hill reaction: The concomitant reduction of an electron carrier and oxidation of water to molecular oxygen that is performed by illuminated green plants; the *light reaction* of photosynthesis. (*see also* Calvin cycle (reductive pentose cycle); Z-scheme)

hinge: A flexible polypeptide sequence connecting two domains of a protein that can move with respect to each other

Hirt lysate: A preparation from which low-molecular-mass extrachromosomal DNA may be isolated; made by rendering insoluble the higher-molecular-mass DNA and other cellular structures by treatment of cells with a detergent and a concentrated salt solution.

histone: One of five small, basic, proteins that are incorporated into chromatin. (*see also* nucleosome)

HLA: (= human leucocyte antigen (HLA))

HLH protein: One of a group of proteins defined by their characteristic topography (the helix-loop-helix). These proteins form DNA-binding heterodimers composed of two proteins, one from a ubiquitous and one from a tissue-specific class of HLH proteins. The heterodimers are involved in regulation of gene expression. (*see also* helix-turn-helix (HTH) motif)

HLH: (= helix-loop-helix (HLH))

HMG protein: (= high-mobility-group (HMG) protein)

hnRNA: Heterogeneous nuclear RNA. (= pre-mRNA (heterogeneous nuclear mRNA))

Hofmeister series: (= lyotropic series (Hofmeister series))

Hogness box: (= TATA box)

Holliday junction: (*see* double-strand-break repair model; Holliday model; *see* Meselson-Radding model)

Holliday model: A conceptualization of general genetic recombination of newly replicated DNA during meiosis. An endonuclease nicks one strand of each of the two daughter duplexes at homologous sites which permits them to cross-over before being ligated. *Branch migration* of the crossing-over point, the *Holliday junction*, along the strands is also possible before they separate to form the recombinant products; *chi-forms* (χ-shaped structures) are generated as the strands separate. In replication of closed circular (cc)DNA, *figure-of-eight forms* are generated as the two daughter ccDNAs separate, except where they have crossed-over and still adhere to each other. (*see also* double-strand-break repair model; Meselson-Radding model)

holocrine: Descriptive of a secretion mechanism in which a cell releases its product by rupture and death of the entire cell. (*see also* apocrine; exocrine)

holoprotein: The complex of an apoprotein with its prosthetic group or metal ion.

homeobox: A highly conserved 180-base polynucleotide sequence that controls body part-, organ- or tissue-specific gene expression, e.g. development of antennae or legs of *Drosophila*, but also in a wide variety of other eukaryotes. It codes for a helix-turn-helix DNA-binding region, a *homeodomain*, of proteins that are transcription factors.

homeodomain: (*see* homeobox)

homeostasis: (= homoeostasis)

homing: The migration of lymphocytes to areas of inflammation. (*see also* intron homing)

homo-: A prefix derived from the Greek for same; as opposed to hetero- from the Greek for other.

homochromatography: A form of displacement chromatography in which unlabelled species, e.g. oligonucleotides, displace labelled ones to achieve improved separation. After electrophoresis in one dimension the initially separated material is transferred from a paper strip to a thin layer chromatography plate that is developed with a solution that contains the unlabelled species and displaces the labelled ones from sites on the ion-exchange stationary phase.

homoeostasis: Self-regulation by an organism to maintain a relatively constant internal environment.

homogenate: An unfractionated, cell-free, preparation of tissue prepared by disruption of tissue structure and breakage of cell walls, e.g. with a Potter-Elvehjem homogenizer or by use of ultrasonic vibrations.

homogeneous: *see* heterogeneous

homoiothermic: Descriptive of an organism that can maintain a fixed internal temperature. (*see also* poikilothermic)

homolactate fermentation: The fermentation of a hexose that produces only lactic acid, as opposed to *heterolactate fermentation*, which also produces other products.

homologous ligation: A method for fusion of two oligonucleotides, independently of any restriction site they share. A region flanking the desired ligation site on one fragment is introduced into the second fragment by PCR as part of one of the primers. The first fragment and the amplified second fragment are introduced in tandem into an appropriate plasmid that can be cleaved by a restriction nuclease engineered between the two fragments. The linearized DNA is used to transform a bacterium that has a high frequency of intramolecular homologous recombinations. Subsequent homologous recombination fuses the two fragments at their homologous region.

homologous recombination hot spot: A DNA sequence that, by providing a specific protein- and RNA-binding site, shows a high frequency of homologous recombination events.

homologous recombination: The non-symmetrical crossing-over of genes at homologous sequences (e.g. at *Alu* sequences of introns between different exons of a single gene) which gives rise to two mutant alleles, one a deletion and the other an insertion; recognized in haemoglobin and low-density-lipoprotein receptor mutations. (*see also* illegitimate recombination)

homologous replacement: A method to alter a cell's genome by introduction of a vector into the nucleus of a targeted germ or somatic cell, where it can pair with the cognate sequence of the chromosomal DNA and replace the homologous region.

homologue cloning: A strategy for selection of polynucleotide sequences closely homologous to one that is known by its partial hybridization with a probe designed for the latter; e.g. a labelled probe that is complementary to a polynucleotide sequence that encodes one histamine receptor may hybridize with a homologous sequence that encodes another histamine receptor under non-stringent hybridizing conditions and allow its identification and selection. (*see also* stringency)

homologue-scanning mutagenesis: A strategy for identification of receptor-binding regions of a protein by substitution of analogous regions from homologous proteins in order to preserve the native three-dimensional structure of the original protein; e.g. the substitution of regions of human growth hormone with regions from pig growth hormone, human prolactin or human placental lactogen, followed by determination of binding constants for the constructs.

homology: The similarity in base sequences of genes or amino acid sequences of proteins that denotes a common evolutionary origin; also the similarity of structure or function of proteins that is due to a common evolutionary origin.

homolytic cleavage: The splitting of a covalent bond that leaves one of the bonding electrons with each of the atoms, thus generating free radicals. (*see also* heterolytic cleavage)

homomer: A complex composed of only one kind of subunit; e.g. the neo-natal glycine-gated Cl^- channel, which is composed of five $_2$ subunits. This is contrasted with a heteromer, a complex composed of more than one kind of subunit, e.g. the adult glycine-gated Cl^- channel, which is composed of three $_1$ and two subunits.

homomeric: Descriptive of a complex of a single kind of subunit, e.g. bacterial glutamine synthetase; contrasted with *heteromeric*, which is descriptive of a complex of two or more kinds of subunit, e.g. haemoglobin A.

homoplasy: In phylogenetic taxonomy, a non-inherited characteristic shared by a group that itself may have a common ancestor.

homopolymer tailing: (= tailing)

homotropic enzyme: An enzyme that is controlled by an allosteric activator that is also a substrate. (*see also* heterotropic enzyme)

homozygote: (*see* allele)

Hoogstein base pairs: The unique pairing that occurs when a homopyrimidine strand, which lies in the major groove of double-stranded DNA, interacts with a sequence of a Watson-Crick base-paired double helix, resulting in a triple helix. Such pairing is possible when the original double helix features a homopurine-to-homopyrimidine base-paired sequence. Allowed hydrogen-bonded base pairs are thymine to an adenine- thymine pair, or cytosine to a guanine-cytosine pair.

hook region: A normally protected but reactive sequence of the 2-macroglobulin molecule, which is a scavenger for many proteinases in the blood. A proteinase can cleave a peptide bond in the protein's bait region, but then becomes susceptible to reaction with the thioester bond formed between a -carboxy group and a -thiol group. Once caught by this hook, the proteinase remains covalently bound to the protein, and is cleared with it from the circulation.

hopanoid: One of a class of pentacyclic triterpenes derived from squalene that occur in the membranes of bacteria and some plants, fulfilling the functions served by sterols in the membranes of other organisms.

hormone: A substance that mediates interactions between non-contiguous cells; classically, a substance produced in minute amounts by an organ (an endocrine organ) in one part of the body and transported by the blood to a distant target organ, which it stimulates; less rigorously applied to substances that share some part of the classical description.

host-guest protocol: An experiment designed to evaluate the contribution of different amino acids at a single sequence position to the stability of a polypeptide. Site-directed mutagenesis or synthetic chemical techniques are used to construct polypeptides that differ only at a single sequence position. Quantification of the stability of the polypeptide measures the contribution of each possible amino acid.

hot spot: A region of a polynucleotide that experiences a high frequency of mutation. (*see also* homologous recombination hot spot)

hot start PCR: A variation of the basic PCR procedure that prevents copying of non-specific primer-template complexes, which may be stable at moderate temperatures, by addition of an essential reaction component only when the temperature has reached 958C during the first cycle.

HOT technique: (= hydroxylamine and osmium tetroxide (HOT) technique (chemical mismatch detection))

housecleaning enzyme: An enzyme that hydrolyses or otherwise destroys a potentially mutagenic or physiologically disruptive metabolite; e.g. a nudix hydrolase, an enzyme that hydrolyzes a nucleoside diphosphatase-linked metabolite.

housekeeping: Descriptive of a gene or enzyme that serves an essential function common to all or most cells, e.g. production of energy; as distinguished from specialized functions, such as differentiation or apoptosis.

HpaII tiny fragment: (= HTF island)

HPLC: (= high-pressure (high-performance) liquid chromatography (HPLC))

HSE: (= heat-shock consensus element (HSE))

Hsp: A prefix which denotes a heat-shock protein; e.g. Hsp60 is a 60-KDa heat-shock protein. (*see* heat-shock protein)

HTF island: A tiny fragment, 1000-2000bp long, generated by action of the restriction nuclease *Hpa*II; usually found associated with expressed genes; characterized by the relatively rare CpG dinucleotide that occurs unmethylated.

HTH: (= helix-turn-helix (HTH) motif)

human artificial chromosome (HAC): An artificial assembly of chromosome features, telomere, centromere, etc., that is used to test the minimum features required for replication in transfected cells. The mammalian artificial chromosome (MAC) serves a parallel function.

human leucocyte antigen (HLA): An antigen on the surface of human cells that consists of a generally invariant $_2$-microglobulin and a larger polypeptide chain that is characteristic of the individual; responsible for most cases of organ transplant rejection. (*see also* major histocompatibility complex (MHC))

humanized antibody: A chimaeric immunoglobulin of human and non-human origin. In order to minimize the possibility of the chimaera being antigenic in humans, only the hypervariable regions of a non-human monoclonal antibody, which are responsible for its specificity, replace the homologous regions of a human immunoglobulin.

Hunter and Downs plot: A graphical method for determination of dissociation constants of enzyme inhibitors. Where I is total inhibitor concentration and α is the ratio of inhibited to non-inhibited rates, a plot of $I\alpha/(1\text{-}\alpha)$ on the ordinate (y-axis) against substrate concentration on the abscissa (x-axis) will give a line that intersects the ordinate at K_i; if inhibition is competitive, the slope of the plot is K_i/K_m, if non-competitive, the slope is zero. The advantage of this method lies in its not requiring that data be collected at constant substrate or inhibitor concentration. (*see also* Cornish-Bowden plot; Dixon plot; Easson and Stedman plot)

hybrid hybridoma: A cell that produces antibodies with dual specificity because it is a fusion of two hybridomas, and therefore produces immunoglobulins characteristic of each parent cell and hybrids that display one binding site of each parental type.

hybrid: In nucleic acid chemistry, a double-stranded polynucleotide, one strand being an RNA and the other a DNA.

hybridization probe: A polynucleotide, often radiolabelled, used to detect complementary sequences, e.g. an mRNA used to locate its gene by a corresponding Southern blotting method.

hybridization: (= annealing (hybridization))

hybridoma: A cell that is useful in production of quantities of a monoclonal antibody; a fusion of a myeloma cell and a spleen lymphocyte that, like the myeloma cell, can divide in culture and, like the spleen cell, can produce antibodies. The use of HAT (hypoxanthine/ aminopterin/thymidine) medium allows selection of hybrids of myeloma cell mutants that lack hypoxanthine-guanine phosphoribosyltransferase and consequently cannot utilize hypoxanthine, with spleen cells, which do not proliferate *in vitro*. (*see also* hypoxanthine/aminopterin/thymidine (HAT) medium)

hybrid-type carbohydrate: One type of glycoprotein moiety that is attached to the -amide nitrogen of an asparagine residue (N-linked) and contains both mannose and sialic acid residues at the non-reducing ends. (*see also* complex-type carbohydrate; high-mannose-type carbohydrate)

hydrazinolysis: In determination of the C-terminal amino acid of a protein, cleavage of a peptide bond by insertion across it of the elements of hydrazine; in the Maxam-Gilbert method for sequencing polynucleotides, the use of hydrazine to convert pyrimidine bases into pyrazoles and remove them from the nucleic acid backbone.

hydro- : A prefix derived from the Greek for water which refers to water as a molecule, to bulk water or to a property displayed in aqueous solution.

hydrodynamic volume: The volume of a macromolecule as deduced from its behaviour in solution. (*see also* Stokes radius)

hydrogen bond: An important cohesive force of biological macromolecules, notably interstrand interactions of bases in the Watson-Crick model of DNA and of amide groups in the -helix of proteins. The bond is the attraction of a weak acid to a weak base due to the sharing of a proton between them, e.g. -OH···O=, ;NH···O=. The sharing may be considered to be an equilibrium between two tautomeric forms, e.g. $-O^{-}\cdots HO^{+}=$, each form lying in an energy well with an energy barrier between them (the *double-well hydrogen bond*). If water is excluded from the area, and if the base strengths of the two atoms that share the proton are equal, the distance between the two nucleophilic atoms is shorter and the bond is much stronger [;100kJ/mol (24kcal/mol); the *single-well hydrogen bond*]. An intermediate case pertains [50-100kJ/mol (12-24 kcal/mol); the *low-barrier hydrogen bond*] when the pK_a values of the nucleophiles are similar and water is largely excluded, as at the active site of an enzyme in the transition state with its substrate.

hydrogen exchange: A method for evaluating secondary and tertiary structure of a polypeptide in solution. The amide hydrogens of the polypeptide backbone slowly exchange with water hydrogens, unless they are non-covalently bonded to neighboring basic groups, such as amide oxygen atoms (as in an a-helix) or carboxylates groups. If the peptide is first equilibrated with D_2O to replace exchangeable hydrogens with deuterium, and then placed in H_2O, the rate of appearance of deuterium in the bulk water phase will be determined by the chemistry of the sites to which deuterium is bound (aminé hydrogens will very rapidly exchange) and availability of the exchangeable sites, as determined by the secondary or tertiary structure.

hydrogenation: The reverse of dehydrogenation.

hydrolase: One of the class of enzymes that transfer a chemical group from donor substrates to water, e.g. peptidase, nuclease, glycosylase.

hydron: A ^1H, ^2H (deuterium) or ^3H (tritium) atom.

hydropathy index: A measure of polarity of an amino acid residue; the free energy of transfer of the residue from a medium of low dielectric constant to water. (*see also* optimal matching hydrophobicity (OMH))

hydropathy plot: A graph that shows regions of hydrophobicity or polarity of a protein; a line graph that displays the hydropathy index against the position of each residue of the protein. Each residue is averaged with adjacent residues in the priary sequence to smooth the plot and allow an evaluation of regions of hydrophobic residues that may form the interiors of globular proteins or transmembrane domains of membrane bound proteins. (*see also* optimal matching hydrophobicity (OMH))

hydrophilic residue: An amino acid residue that has a charged or polar side chain.

hydrophobic bond: The association of non-polar side chains of a protein that is driven by minimization of the relatively unfavourable interactions of water molecules with the non-polar groups, and maximization of the favourable interaction of water molecules with themselves when they are removed from contact with the non-polar residues and are added to the bulk water phase.

hydrophobic cluster analysis (HCA): A method for comparison of amino acid sequences of proteins to detect regions of conformational similarity. Groups of neighbouring hydrophobic amino acid residues are identified in a HCA plot that is constructed by conceptually folding the entire backbone into an -helix, rolling the helix two complete revolutions across a two-dimensional surface and noting the positions where -carbons have contacted the surface. Each contact point is indicated according to the residues attached to it. The two-dimensional visual pattern of neighbouring hydrophobic residues, whether adjacent in the primary sequence or brought together by folding of the helix, is characteristic of the actual conformation, and serves as a basis for comparison of proteins. (*see also* structural profile)

hydrophobic collapse: The initial stages in protein folding, in which interactions of local hydrophobic groups interact and quickly reduce the size of the unfolded protein. (*see* hydrophobic zipping)

hydrophobic moment: A measure of the amphipathicity of an -helix. Each amino acid side chain is assigned a value, positive or negative, according to its hydrophobicity. These values are used to weight vectors for each residue as they are displayed around the helix. The summation of the vectors is the hydrophobic moment, ;$_M$;. Mean *hydrophobicity*, ;H; is the sum of the hydrophobicity values divided by the number of residues. (*see also* Eisenberg plot)

hydrophobic zipper: A feature of the structure of some proteins, in which neighbouring hydrophobic residues of parallel or antiparallel -strands interdigitate and form a hydrophobic cluster.

hydrophobic zipping: A view of the initiation of protein folding in which the formation of hydrophobic interactions between very close groups bring slightly more remote groups into interaction distance in an accelerating condensation.

hydrophobicity scale: A ranking of amino acid residues or side chains according to some quantitative measure of their hydrophobicity, e.g. the free energy of transfer of the side chain from dioxane or an alcohol to water.

hydrophobicity: (*see* hydrophobic moment)

hydroxyl radical footprinting: The use of pulses of X-rays from a synchrotron to generate hydroxyl radicals, which will quickly react with, and thus label exposed residues of, a macromolecule as time intervals as short as 10 msec. A large unfolded tRNA of *Tetrahymena*, for instance, may be induced to begin its folding by the addition of Mg^{2+}, while at timed intervals, bursts of hydroxyl radicals label the exposed regions of the RNA.

hydroxylamine and osmium tetroxide (HOT) technique (chemical mismatch detection): A method for detection of single base substitutions in DNA fragments. Heteroduplexes of normal DNA and homologous mutant DNA are generated; the presence of C mismatches allows cleavage with hydroxylamine and the presence of C and T mismatches allows cleavage with osmium tetroxide. Cleavages are detected after polyacrylamide-gel electrophoresis.

hyper-: A prefix derived from the Greek for over or beyond which indicates an excess. *see* hypo-

hyperchromic effect: The increase in absorbance, frequently measured at 260nm, when a native polynucleotide is denatured. The ratio of absorbance of the native polymer to that of the denatured or hydrolysed material is the *hyperchromicity ratio*, and ranges from 1.0 to 1.4. (*see also* hypochromicity)

hyperchromicity ratio: (*see* hyperchromic effect)

hyperfine splitting: The detail of a nuclear magnetic resonance spectrum, due to interactions of adjacent nuclei, that is superimposed on the Zeeman splitting.

hyperplasia: An excessive growth of an organ due to an increase in the number of cells. (*see also* hypertrophy)

hypertensive: Descriptive of the action of a substance in raising blood pressure. (*see also* hypotensive)

hyperthermophile: (*see* extremophile)

hypertonic: (*see* osmolarity)

hypertrophy: An excessive development of an organ, especially due to overstimulation, but without an increase in cell number. (*see also* atrophy; hyperplasia)

hypervariable epitope construct: (*see* mixotope)

hypervariable region: A localized sequence of a variable region of an immunoglobulin that shows a particularly large variation in amino acid sequence compared with all other immunoglobulins; presumably the part of the immunoglobulin that interacts directly with the antigen. (*see also* constant region; variable region)

hypo-: A prefix derived from the Greek for under which indicates a dearth. (*see* hyper-)

hypochromicity: The decrease in absorbance at 260nm observed when a strand of single-stranded DNA base-pairs with its complementary strand to form double-stranded DNA. (*see also* hyperchromic effect)

hypoglycaemic (hypoglycemic) effect: The action of insulin in lowering blood glucose levels by making adipocytes and skeletal muscle cells permeable to the sugar.

hypotensive: Descriptive of the action of a substance in lowering blood pressure. (*see also* hypertensive)

hypotonic: (*see* osmolarity)

hypoxanthine/aminopterin/thymidine (HAT) medium: A growth medium that is used to isolate monoclonal-antibody-producing cells, i.e. hybrids made from fusion of myeloma cell mutants that lack hypoxanthine-guanine phosphoribosyltransferase, and consequently cannot utilize hypoxanthine, with spleen cells, which do not proliferate *in vitro*. (*see also* hybridoma)

H-zone: The central area of a sarcomere, identified microscopically; traversed by thick filaments but not thin filaments.

I

idio-: A prefix derived from the Greek for own, or self which indicates uniqueness.

idiotype: A classification of immunoglobulin molecules according to the antigenicity of the variable regions. Each idiotype is unique to a particular immunoglobulin raised to a particular antigen. (*see also* allotype; anti-ergotypic; isotype)

IEF: (= isoelectric focusing (IEF))

illegitimate recombination: Recombination with non-homologous regions of DNA. Each of the polynucleotide strands of double-stranded DNA is religated with DNA from different regions of the same or different chromosomes. (*see also* homologous recombination)

image plate: A kind of area detector; a phosphor-impregnated plate that can be scanned to quantify its radiation exposure, and can then be regenerated.

imino group: A secondary amine, e.g. the ;NH group of proline.

iminum ion: (*see* a-type ion)

immediate-early stage: The first stage of response of a cell to viral infection, followed by the *delayed-early stage* and the *late stage*. Each stage is characterized by transcription of specific viral genes.

immobile DNA junction: A branched double-stranded DNA structure, a chi-form in which each of four polynucleotide chains is complementary to and base-paired with the 3'-end of one polynucleotide and the 5'-end of another; unlike in a recombination intermediate, the junction cannot migrate (i.e. slide its branch point along the duplex) without breaking many more Watson-Crick base-pair hydrogen bonds as it advances than can re-form behind it. (*see also* Holliday model)

immunoaffinity chromatography: A variant of affinity chromatography in which an antibody coupled to the stationary phase adsorbs an antigen which is subsequently eluted at a different pH or a higher salt concentration. Specific antibodies may similarly be prepared on columns of immobilized antigens.

immunoblotting: (= western blotting (immunoblotting))

immunodiffusion: A method for testing the identity of several antigens using an *Ouchterlony plate*, a glass plate covered with a thin layer of agar on which is incised a hole surrounded by several other holes that serve as wells. In the central well is placed an antibody, and the antigens in question are placed in the surrounding wells; the antibody and antigens diffuse towards each other until they meet and form a visible precipitin line. An antigen in an adjacent well forms a precipitin line that is continuous with the first only if the two antigens are identical. Similarly, the plate can be used to test the identity of antibodies against a single antigen in the central well.

immunoelectrophoresis: A method for analysis of antibody or antigen mixtures. A solution of antibodies is electrophoresed in a gel, then exposed to an antigen solution placed in a trough cut into the gel parallel to the direction of the separation. The antibodies and antigens diffuse into the gel, and where they meet they precipitate and form a visual or stainable arc.

immunofluorescence: The visualization of an antigen in a histological section or on a polyacrylamide slab by allowing it to combine with a fluorophore-tagged antibody. In *indirect immunofluorescence* the antibody is not itself labelled but is the target of another antibody which is labelled, e.g. detection of an antigen by a specific human antibody that depends upon the subsequent attachment to the antigen-(human antibody) complex of a rabbit antibody directed against a broad range of human immunoglobulins.

immunogen: (= antigen (immunogen))

immunoglobulin: One of a member of five general classes of serum and blood cell proteins that recognize foreign compounds.

immunoglobulin fold: The characteristic supersecondary structure of each of the domains of immunoglobulins that consists of seven -pleated sheets in two antiparallel layers.

immunomics: (*see* omic research)

immuno-PCR: A very sensitive method for detection of antigens. The antigen is immobilized in wells of a microtitre plate, exposed to its specific monoclonal antibody, then to a chimaeric protein that consists of a protein A moiety, which will bind tightly to the immunoglobulin G antibody, and a streptavidin moiety, which will bind to a biotinylated DNA fragment. The fragment is amplified by PCR, separated by gel electrophoresis and stained with ethidium bromide. The chimaeric protein, biotinylated DNA fragment and PCR primers may be used to assay any antigen-antibody pair. The system is sensitive to as few as 580 antigen molecules (10^{-21} mol).

immunoprecipitation: A purification technique that separates antigenic material from a soluble mixture by precipitation with an appropriate antibody; an essential step in radioimmunoassays.

immunotoxin (chimaeric toxin): A conjugate of an immunoglobulin and a toxin which is designed to mimic a type 2 ribosome-inactivating protein; the immunoglobulin recognizes a target cell and the toxin inactivates it.

imposter: A gene product that can replace a missing homologous protein. Knockout experiments are sometimes misleading because a physiological function is preserved by an imposter; in such cases, a lethal mutation can produce a more informative phenoytpe than the knockout.

imprinting: The process by which some genes are rendered non-equivalent. The paternal or maternal allele is not expressed (allelic exclusion), or is expressed differently in different tissues. The phenomenon is possibly directed by an imprinting box, a parent-specific polynucleotide sequence that instructs an imprinting factor, which may act by methlyation of differentially methylated regions (DMRs), and which will reversibly activate or inactivate the gene inherited from only one parent. Imprinting passes through erasure of methylation of both parental and maternal chromosomes during germ cell development, establishment of methylation, according to the sex of the gamete, and after fertilization, maintenance of the methylation status of the genome during the mitoses of embryogenesis. (*see* DNA methylation)

in silico: *Descriptive of research into bioinformatics databases.*

in situ: 'In place'; descriptive of, for example, perfusion of the liver within the abdominal cavity rather than excision of the organ and perfusion *in vitro.*

in situ PCR: The performance of PCR on fixed preparations on a microscope slide.

in vitro: 'In glass'; taken into an experimental situation, as opposed to *in situ* or *in vivo.*

in vitro evolution: The optimization of a gene product function (binding or catalysis) by an iterative procedure involving mutation, expression, selection and replication. In at least one example, continuous *in vitro* evolution has been achieved in optimizing the condensation of an oligodeoxynucleotide with a promoter.

in vitro selection: (= systematic evolution of ligands by exponential enrichment; *see* cyclic amplification and selection of targets (CASTing))

in vitro synthesized protein assay (IVSP): *see* protein truncation test

in vivo: In the living organism, as opposed to *in vitro*. In biochemistry the term often refers to activity within a living cell.

inborn error: An inherited defect in a metabolic enzyme or other gene product.

incision: (*see* excision)

included mutation: A mutation that is expressed in a protein that is itself a subunit of a larger complex.

inclusion body: An amorphous deposit in the cytoplasm of a cell; an aggregated protein appropriate to the cell but damaged, improperly folded or liganded, or a similarly inappropriately processed foreign protein, such as a viral coat protein or recombinant DNA product. (*see also* chaperone machine; Heinz body)

indel: = signature sequence

indirect bilirubin: (*see* van den Bergh reaction)

indirect calorimetry: Evaluation of the heat evolved by a subject by measurement of oxygen consumption, carbon dioxide evolution and excretion of nitrogenous metabolites. (*see also* direct calorimetry)

indirect end labelling: A method to locate, between restriction sites, binding sites on DNA, e.g. the positions of nucleosomes on a minichromosome. The DNA protected by binding proteins is cleaved in susceptible, i.e. unprotected, regions by non-specific chemical or enzymic methods, the protecting protein is removed, and a restriction nuclease introduces reference points recognized by complementary polynucleotide probes for subsequent analysis of the sequences from the restriction sites towards the sites at which it was initially cleaved in the unprotected regions.

indirect van den Bergh reaction: (*see* van den Bergh reaction)

induced-fit theory (rack model): A model for interaction of a protein with a ligand in which the protein binds preferentially with one of several conformers of the ligand; especially enzyme action in which the substrate is proposed to bind to a conformation of the enzyme that produces a strain on the substrate, focused on the bond that must be broken to effect catalysis; also known as the *handshake model*, and contrasted with the *lock and key theory*, in which both protein and ligand are imagined to be inflexible.

inducer: (*see* induction)

inducible enzyme: An enzyme whose rate of synthesis can be controlled by an inducer. (*see also* constitutive enzyme; induction)

induction: The production of an enzyme in response to the presence of a particular compound, the inducer, or a condition, e.g. heat. In developmental biology, induction is the effect of one group of cells upon the development of another group, mediated by a chemical property, e.g. a cell surface protein, or a secretion of the former group.

informational RNA (iRNA): Essentially, intronic sequences of hnRNA, for which an evolutionary function has been proposed: by a mechanism analogous to the SELEX procedure, and in the absence of any conservative selective pressure, intronic RNA is recruited for evolution of new genes. (*see* cyclic amplification and selection of targets)

infra-: A prefix derived from the Latin for below.

Infrared spectroscopy: IR spectroscopy; a technique that measures absorption of electromagnetic radiation between the long wavelength end of the visible spectrum (780-800 nm) and the short end of the microwave spectrum (1000 m). In practice, most commercial IR spectrometers measure from 780 nm to 300 μm. Absorption in the IR is due to interatomic motions of molecules, vibrations and rotations about bonds. Spectra are commonly presented as percent transmission vs. wavelength. (*see* Fourier transform infrared spectroscopy (FT-IR))

inhibitor: In enzymology, a compound, or even a macromolecule, that blocks the action of an enzyme by reversible attachment in such a way as to prevent binding by the substrate (*competitive inhibition*), or by prevention of the reaction even if the substrate can still bind (*non-competitive inhibition*).

initial velocity: The rate of an enzymic or chemical reaction as it begins, i.e. with all concentrations of substrates and products at defined levels.

initiation codon: A trinucleotide sequence, AUG, that signals the start of translation of a protein; the codon for methionine in eukaryotes and for *N*-formylmethionine in prokaryotes

initiation complex: The ribosome charged with mRNA, initiation factors and methionyl-tRNA (in eukaryotes) or *N*-formylmethionyl-tRNA (in prokaryotes, mitochondria and chloroplasts) that is bound to the initiation codon. (*see also* pre-initiation complex)

initiation factor: An accessory protein that is necessary for assembly of the ribosome-mRNA complex and the start of protein synthesis.

initiator tRNA: The tRNA that recognizes the AUG start codon. In prokaryotes it can be charged to bear a formylmethionine residue (methionine in eukaryotes) which will become the N-terminus of the new protein.

initiator: A site, upstream from a structural gene, for attachment of a protein that stimulates initiation of transcription. (*see also* promoter)

innate immunity: A system of blood factors, especially the complement system, and cells, especially macrophages, which affords protection against invading micro-organisms. Unlike acquired or adaptive immunity, which is based upon recognition by antibodies, innate immunity is independent of prior exposure and depends upon pattern-recognition receptors of relatively broad specificity, particularly those on the surface of macrophages, which allow recognition of micro-organisms and opsonized particles.

inner filter effect: A fluorescence spectroscopy phenomenon; the decrease in fluorescence emission seen in concentrated solutions due to the absorption of exciting light by the fluorophore that is close to the incident beam and which significantly diminishes light that reaches the sample further away from it.

insertion mutation: A mutation caused by the presence of one or several extra nucleotide residues in the DNA sequence, which may result in a frameshift.

insertion sequence: A small transposon, bordered at its 3'- and 5'-ends by inverted terminal repeats.

instructive theory: A now discarded hypothesis that explains antibody specificity by the folding of an antibody structure, which has the potential for many specificities, to the unique surface of a particular antigen. (*see also* selective theory (clonal selection theory))

insulator boundary element: A sequence in genomic DNA which sets a limit to heterochromatin structure.

insulator: A DNA sequence that contributes to limitation of chromatin activation to distinct segments. In the domain boundary model, the insulator putatively binds nucleoprotein complexes along the chromatin strand and, by drawing together these boundary complexes, simultaneously condenses the chromatin and physically separates enhancer and promoter sequences, thus preventing activation of genes in this region. In the transcriptional decoy model, the insulator is imagined to provides a site for assembly of a decoy complex with which the enhancer will interact in preference to, or in competition with, the true promoter.

intasome: A nucleoprotein complex active in integration of bacteriophage DNA into host DNA. Negatively supercoiled host DNA is forced to cross back over itself at points called *nodes*, which are sites for binding of proteins that facilitate the recombination with phage DNA or result in crossing-over of the host strands and formation of a knotted DNA.

integral protein (intrinsic protein): A protein intimately embedded in a membrane, frequently even passing through it and emerging on both sides. (*see also* peripheral protein (extrinsic protein))

intein: The protein counterpart of the RNA intron; a polypeptide sequence that is excised by a self-catalysing mechanism from the primary translation product. (*see also* protein splicing)

inter-: A prefix derived from the Latin for between or among.

intercalating agent: A compound that acts by inserting itself between adjacent bases in a DNA chain; during replication it is capable of causing either a deletion or an insertion mutation.

interchromomere: (*see* AT queue)

interference optics: (*see* Fourier transform infrared spectroscopy)

interferon: One of a group of small anti-viral proteins synthesized and secreted by cells following viral infection.

interleukin-converting enzyme (ICE): (*see* caspase)

intermediary metabolism: The individual enzymic reactions that in a tissue or inside a cell transform one metabolite into another, e.g. the conversion of lactate into glucose by the liver, the conversion of leucine into acetoacetate by muscle, the conversion of sucrose into ethanol by yeast.

intermediate filament: A structure intermediate in diameter between a thin filament and a microtubule, usually 7-11nm in diameter; a neurofilament, keratin molecule, etc.

internal control region: A regulatory site of a gene that interrupts a structural gene.

internal guide sequence: A polynucleotide sequence near the 5'-end of group I introns that pairs with sequences of the upstream exon in an intermediate of the self-splicing process. (*see also* self-splicing)

internal ribosome entry site (IRES): In the 5'-cap-independent initiation of translation, the place on the mRNA that first attaches a ribosome subunit. (*see* ribosome jumping)

internal signal sequence: An internal polypeptide sequence of a protein that is responsible for its targeting to the appropriate locus within a cell after synthesis. (*see also* signal sequence (leader sequence))

interphase nucleus mapping: A variant of the usual fluorescence *in situ* hybridization technique in which non-mitosing cells are used to visualize a genetic marker on a (non-visible) chromosome by use of a fluorescence-labelled polynucleotide probe that hybridizes to and indicates the locus of a gene on a chromosome.

inter-resource duplex: (*see* subtractive DNA cloning)

interspersed repetitive sequence-PCR (IRS-PCR): A method for amplification of human DNA sequences in hybrid somatic cells, using primers directed at species-specific repetitive sequences: a short interspersed repeat element (e.g. *Alu*) sequence, a long interspersed repeat sequence or a combination of both.

intervening sequence region: (= intron)

intimacy model: (*see* altered-self hypothesis)

intra-: A prefix derived from the Latin for inside.

intrinsic factor: A glycoprotein secreted by the gastric mucosa that assists in the absorption of vitamin B_{12} (extrinsic factor) in the intestine; absent from subjects with pernicious anaemia.

intrinsic pathway: The blood clotting cascade that is initiated by a factor that can be generated within the blood itself. (*see also* extrinsic pathway)

intrinsic protein: (= integral protein (intrinsic protein))

intron: From *intervening sequence region*; originally defined as a non-coding polynucleotide sequence that interrupts the coding sequences, the *exons*, of a gene. More recent evidence indicates that, in some cases, introns may also be translated, so a more cautious definition of an intron is merely a DNA sequence whose transcribed sequence is physically separated from that of exons during RNA processing. (*see also* spliceosome)

intron editing: A change in the reading of the sequence coded by DNA directed by an adjacent intron. The intron, which has a sequence complementary to part of the exon, forms a double-stranded stem structure that is subject to enzymic alteration of a coding base, e.g. deamination of an adenine, before the intron is eliminated by splicing. (*see also* spliceosome)

intron homing: The direction of introns, as mobile genetic elements, to sites in DNA. The process is dependent upon proteins that are encoded by open reading frames of the introns: endonucleases in the case of group I introns and reverse transcriptases in the case of group II introns. (*see also* protein splicing)

intron-early: (*see* shuffling)

intron-late: (*see* shuffling)

inverse PCR: A PCR protocol for amplification of sequences that flank a single known sequence. The source DNA is digested with a restriction endonuclease, circularized and ligated. Primers are constructed that will hybridize to the ends of the known sequence, but will be extended in opposite directions into the flanking sequences and around the circle. Cycles of polymerization, denaturation and rehybridization amplify the sequence between the restriction sites. (*see also* ligand-mediated PCR (LMPCR); vectorette PCR)

invertasome: A nucleoprotein complex that mediates inversion of a DNA sequence, bracketed by two recombination sites, within the genome of an organism.

inverted terminal repeat: A polynucleotide sequence that is repeated in reverse sequence; the flanking sequence of a transposon.

iodine number: A measure of the unsaturation of a lipid; the number of grams of iodine that react with 100g of an olefin.

iodine test: The reaction of some polysaccharides, e.g. starch or glycogen, with iodine to give a red to purple colour; diagnostic of an a-linked polysaccharide.

ion trap: A mass spectrograph in which vaporized ions are confined by an AC electric field. As the radio-frequency voltage that confines the ion is changed, the ion will escape the trap at a voltage characteristic of its mass-to-charge ratio, and be detected. Sequences of polypeptides and polynucleotides may be determined by expulsion from the trap of all but one kind of large mass ion, fragmentation of this ion by collision with helium atoms and analysis of the mass spectrum of the resulting fragments. A variant of this electrical-type ion trap uses a magnetic field to confine ions. (*see also* tandem mass spectrometry (tandem MS; MS/MS))

ionophore: An agent that allows passage of an ion through an otherwise impermeable membrane, e.g. the protonophore thermogenin, which makes the inner membrane of brown adipose tissue mitochondria permeable to protons.

ionotropic: Descriptive of plasma membrane receptors that, when activated, become ion channels; in contrast with metabotropic receptors, the actions of which are mediated by G-proteins.

iontophoresis: The transport of ions across a semi-permeable membrane under the influence of an electric charge. As ions are driven across the membrane by the electric field, there is also the accompanying transport of bulk water (electro-osmosis) and uncharged solutes.

IR spectroscopy: (*see* infrared spectroscopy

IRD: Inter-resource duplex. (*see* subtractive DNA cloning)

iron-sulphur cluster: The complex of iron and sulphur atoms in an iron-sulphur protein.

iron-sulphur protein: One of a class of redox-sensitive proteins and other enzymes characterized by a prosthetic group of one to four iron atoms, depending upon the protein, which are co-ordinated with the sulphur atoms of cysteine and/or with suphide ions; e.g. succinate dehydrogenase, aconitase. The most common examples contain a Fe_2S_2, Fe_3S_4 or Fe_4S_4 (cubane) cluster. (*see* sirohaeme)

IRS-PCR: A method for amplification of human DNA sequences in hybrid somatic cells, using primers directed at species-specific repetitive sequences: a short interspersed repeat element (e.g. *Alu*) sequence, a long interspersed repeat sequence or a combination of both.

ischaemia: An insufficient supply of blood to a part of the body with the resultant deficit of oxygen. Ischaemia is a feature of heart attacks, strokes and several kinds of surgery.

islet: A histological structure; usually the islet of Langerhans in the pancreas which consists of glucagon-secreting -cells and insulin-secreting -cells.

isoacceptor: One of two or more tRNAs that accept the same amino acid.

isobol: (*see* isobologram)

isobolcgram: A representation of enzyme kinetic data as a plot of varying substrate and inhibitor concentrations that give constant activity.

isocratic: At constant concentration.

isoelectric focusing (IEF): A technique to separate proteins by their isoelectric points. Proteins are electrophoresed through a gel that has imposed upon it a pH gradient stabilized by the inclusion of polyampholytes, which themselves have a range of isoelectric points. Each protein migrates until it reaches the pH region of its isoelectric point.

isoelectric point: The pH value at which an amphoteric compound is electrically neutral.

isoenzyme (isozyme): One of two or more structural variants of an enzyme that can occur in the same individual, often relatively specific for a particular tissue, each with unique kinetic characteristics; e.g. the heart (H)- and muscle (M)-type lactate dehydrogenase subunits.

isoform: A variant form of a protein. Isoforms may be encoded by different genes, e.g. the isoenzymes of lactate dehydrogenase, or by a single gene, the transcript of which is alternatively spliced and upon translation, yield isoforms that are transported to characteristic subcellular sites, e.g. the isoforms of mammalian acetylcholinesterase.

isohydric shift: The transformations in the erythrocyte that accommodate the uptake and partial binding of carbon dioxide while maintaining a relatively constant pH.

isomerase: One of a class of enzymes that rearrange the bonds of their substrates, e.g. an epimerase.

isomorphous replacement: An X-ray crystallographic technique to assist in solving structures from a diffraction pattern. One or more heavy-metal atoms are inserted in the crystal (replacement) without disturbing the arrangement of the other atoms of the structure (isomorphous). In *single isomorphous replacement* a single derivative is used to solve a molecular structure; in *multiple isomorphous replacement* several derivatives are used. (*see also* difference Patterson map)

isopeptide bond: An amide bond between amino acids that employs a non-amino or a non-carboxy group, e.g. the -carboxy—amino cross-link of hard fibrin clots.

isoprene rule: The observation, first based on inspection of structures and later on experimentation, that many biological substances (e.g. sterols, tocopherols, rubber) appear to have been assembled from multiple isoprene units. (*see also* terpene)

isoprene unit: The branched five-carbon chain that is a recognizable component of terpenes and other compounds derived from isopentenylpyrophosphate, the biosynthetic isoprene unit. (*see also* terpene)

isopycnic centrifugation: (= density-gradient centrifugation)

isoreceptor: One of two or more homologous plasma membrane or cytosolic receptors that have the same or altered functions; analogous to an isoenzyme.

isosbestic point: In spectroscopy, the wavelength at which two or more components of a solution have the same molar absorption coefficient.

isoschizomers: A pair of restriction endonucleases that recognize the same palindromic sequence, although in some cases one member of a pair may differ from the other in its response to methylation of the recognition sequence, or may cleave a different phosphodiester bond within the target sequences.

isotachophoresis: (*see* electrophoresis)

isotherm: A binding curve; at constant temperature, the concentration-dependence of the binding of one material to another, e.g. a gas at various pressures to a solid surface, an amino acid at various concentrations of a neutral salt to a chromatography matrix.

isothermal titration calorimetry: ITC; A technique for measuring binding affinities (free energy of binding) of molecules in the low millimolar to nanomolar range. The heat absorbed or released during binding under equilibrium conditions is detected. ITC does not require that reactants are labelled, fixed to a surface or separated. The enthalpy and entropies of binding may also be measured.

isotonic: (*see* osmolarity)

isotope: A variant of an atom, chemically identical but with a different atomic mass; often radioactive.

isotope dilution assay: A method to determine the amount of a substance in a mixture by addition to it of a known amount of the same material, isotopically labelled and of known specific radioactivity. The material is subsequently isolated in pure form and its new specific radioactivity is determined. From the degree of dilution and the amount of labelled material added, one may calculate the original concentration.

isotope effect: The effect on kinetics of the substitution of one isotope for another in a bond that must be broken in a rate-limiting step (*primary isotope effect*); especially prominent in the case of cleavage of a bond to a deuterium atom compared with a hydrogen atom. Smaller effects are seen in the case of a substitution of an isotope for a non-reacting atom (*secondary isotope effect*); effects are also seen on parameters other than the rate-limiting step. (*see also* solvent isotope effect)

isotropic: Descriptive of a physical property that is independent of the angle of observation. (*see also* anisotropic)

isotype: An immunoglobulin molecule classified according to the amino acid sequences of the constant regions i.e. α, β, δ, ε, γ and μ heavy chains and κ and λ light chains. (*see also* allotype; idiotype)

isozyme: (= isoenzyme (isozyme))

IVS: Intervening sequence. (= intron)

J

J-chain: A polypeptide found in the dimeric immunoglobulin A and the pentameric immunoglobulin M, involved in joining together the subunits of each multimer.

jumping gene: (= transposon)

jumping library: A tool for the mapping and sequencing of large stretches of a chromosome. Starting at one point in the polynucleotide sequence, a distal point is located without having to first sequence the intervening bases. The double-stranded DNA is divided into large fragments with a rare cutter and one end of each piece is ligated to a selective marker and cyclized, then cut with a frequent cutter. The marker allows isolation of the two ends of the original large piece without the intervening sequence.

jumping PCR: A technique for reconstruction and replication of randomly damaged double-stranded DNA, e.g. that found in mummified tissues or museum samples. In the first PCR cycle each primer is extended until a break in the template is encountered, but in the second cycle the newly synthesized strands find other templates that are intact in the region where the first templates were damaged. Each cycle extends the length of newly synthesized strands until complete strands are formed and PCR can begin its exponential synthesis.

junk DNA: DNA for which a function has yet to be identified. An elaborate nomenclature has been proposed for this DNA: any stretch of DNA is a *nuon*; DNA with an evolutionary potential for development of function is a *potonuon*; DNA that was previously non-functional, or that had a completely different function and has been evolutionarily co-opted for a new function (a process termed *exaptation*), is a *xaptonuon*; and a non-co-opted potonuon is a *naptonuon*.

K

17-ketogenic steroid: A steroid that on oxidation with sodium bismuthate will form a 17-keto (-oxo)-steroid; usually a 17-hydroxy-, 20-oxo- or 17, 20-dihydroxy-steroid.

Ka: (= association constant (K_a))

kallikrein: (*see* kinin; kininogen)

karyo-: A prefix derived from the Greek for kernel which indicates a cell nucleus.

karyokinesis: Division of the nucleus of a cell during mitosis.

karyophile: A substance that can diffuse into and be bound in the nucleus of a cell, presumably due to its affinity for a non-diffusible nuclear compound.

karyotype: The display of chromosomes, arranged from largest to smallest, obtained by cutting out and arranging photomicrographs of chromosomes.

katal: A measure of enzyme activity; the conversion of 1 mol of substrate per s under specified conditions. (*see also* unit)

kbp (kb): Kilobase pairs (kilobases).

kcat inhibitor: (*see* mechanism-based inhibitor)

kcat: (= catalytic rate constant (k_{cat}))

Kd: (= dissociation constant (K_d))

kDa: Kilodalton.

ketal: (*see* hemiketal)

ketimine form: (*see* quinimine form)

ketogenic amino acid: An amino acid whose carbon skeleton can be converted, at least in part, into ketone bodies. (*see also* glucogenic amino acid)

ketone body: Acetoacetate, -hydroxybutyrate or acetone.

ketosis: An accumulation in the blood of ketone bodies, indicative of a metabolic dysfunction such as uncontrolled diabetes mellitus or starvation.

killer T-cell: (*see* T-cell)

kilo-: A prefix (abbreviated K) that denotes 10^3; e.g. kilobase, 1 Kbp, 10^3 base pairs.

kilobase (kb): A measure of DNA chain length; 1000 nucleotide residues.

kin-: A prefix derived from the Greek for motion which also indicates activation.

kinase: An enzyme that uses ATP to phosphorylate a substrate; also, in older literature, an enzyme that activates its substrate, e.g. enterokinase. (*see also* protein kinase C (PKC))

kinesis: (*see* protein kinesis)

kinetic assay: An enzyme-based assay that measures the amount of substrate present by correlation of the rate of reaction with the known dependence of the rate on substrate concentration, usually under first-order conditions. (*see also* end-point assay)

kinetic enrichment: (*see* subtractive DNA cloning)

kinetic partitioning: The hypothesis that in folding of a newly synthesized or renaturing biopolymer, a protein or RNA quickly collapses into condensed structures characterized by a high degree of secondary structure. One of these may be the native conformation (the others are "kinetic traps"), but significant energy barriers isolate these forms. Kinetic partitioning is the hypothetical basis of recognition of proteins by molecular chaperones: quickly after synthesis, a cytosolic protein achieves its native conformation and thus avoids binding to the chaperone, while other proteins, e.g. those with leader sequences, much more slowly fold into native structures, which affords the chaperone the opportunity to bind them. (*see* funnel concept; paradox of Levinthal)

kinetic proofreading: A mechanism for assuring fidelity in directed macromolecular synthesis; e.g. in protein synthesis a well matched anti-codon-codon pair, as opposed to a poorly matched pair, will remain associated until commitment to the next step of synthesis, which is timed by an internal 'clock', the hydrolysis of bound GTP. (*see also* G-protein; double-sieve editing)

kinetic trap: (*see* funnel concept; kinetic partitioning)

kinetics: The study of rates of reactions.

kinetochore: The site on a metaphase chromosome to which chromosomal microtubules attach and in anaphase draw them toward the centrioles.

kinin: A hypotensive plasma peptide that dilates small blood vessels and increases capillary permeability, e.g. bradykinin; formed from its precursor, a *kininogen*, by the proteolytic action of a *kallikrein*.

kininogen: A protein that contains the sequence Arg-Pro-Pro-Gly-Phe-Ser-Pro-Arg which has inhibitory activity against cysteine proteinases. A kininogen is a precursor of a *kinin*, which is a decapeptide or undecapeptide produced by the proteolytic activity on a kininogen by a *kallikrein*. An *L-kininogen* is a low-molecular-mass kininogen and an *H-kininogen* is a high-molecular-mass kininogen, e.g. T-kininogen is one of the H-kininogens in the rat. (*see also* $alpha_{1\ 1}$)-cysteine proteinase inhibitor; cystatin; thiostatin (T-kininogen))

kinomics: (*see* omic research)

kissing complex: The interaction of two RNA molecules by hybridization of complementary sequences from a loop of each.

Klenow fragment: The product of limited proteolysis of *E. coli* DNA polymerase I that retains the polymerase and 3'-to-5' exonuclease (proofreading) activity but is missing the 5'-to-3' nuclease activity.

Klett unit: An arbitrary measure of absorbance used in a now-obsolete optical filter-based colorimeter.

Km: (= Michaelis constant (K_m))

KNF model: (= Koshland model (KNF model))

knock-down: A strategy for down-regulation of expression of a gene by incorportation into the genome an antisense oligodeoxynucleotide or ribozyme sequence that is directed against the targeted gene. (*see* knock-out)

knock-in: An organism in which a foreign or mutated gene replaces the normal gene. Unlike in transgenic techniques, the insertion is site-specific and delivers one copy of the gene. (*see* cassette)

knock-out: An animal or plant from which a certain gene has been deliberately deleted, in order to observe, as a test for the gene's physiological significance, the viability or phenotype of the organism. A knock-out is contrasted to a transgenic, in which a gene from another genomic source has been inserted. (*see* knock-down; knock-in)

knot: (*see* node)

Koshland model (KNF model): A sequential model for the basis of co-operativity in multimeric binding proteins, originally presented as a description of oxygen binding to each of the haem groups of the haemoglobin tetramer. Unlike in the earlier Monod-Wyman-Changeux concerted model, each monomer may assume a low-affinity or a high-affinity conformation. A high-affinity ligand-bound monomer favours the high-affinity conformation of adjacent unbound monomers. In the absence of a ligand (oxygen in the case of haemoglobin) the subunits exist in a low-affinity 'taut' conformation. The ligand may bind to as many as two monomers in this conformation, at which point all four monomers convert from the low-affinity form to the high-affinity 'relaxed' conformation.

Krebs cycle: (= tricarboxylic acid cycle (TCA cycle)

Krebs urea cycle: (= urea cycle (Krebs urea cycle))

Krebs-Kornberg cycle: (= glyoxylate cycle (Krebs-Kornberg cycle))

kringle: A protein folding motif, as visualized in two dimensions, of loops formed by multiply disulphide-bridged sequences of several circulating proteins, e.g. $_2$-macroglobulin, complement component C3, prothrombin, tissue-type plasminogen activator.

Kunkel mutagenesis (dUTP system mutagenesis): A method to construct a site-specific mutated gene and to select it away from the wild-type. A strain of *E. coli* with only weak dUTPase activity has high levels of dUTP and consequently incorporates dUTP into a plasmid DNA in competition with dTTP. This plasmid is used in an *in vitro* system as a template for DNA synthesis with a primer into which the desired mutation has been incorporated. The duplex, containing the wild-type sequence with Us replacing Ts (*U-DNA*) complementary to a strand with the usual DNA bases and incorporating the mutation, is introduced into a wild-type bacterium that can remove the inappropriate bases from, and thus inactivate, the U-DNA strand and leave the mutated strand to be replicated.

L

label: A chemical modification or isotope substitution that makes possible the identification of a molecule, moiety or atom throughout its metabolic transformations or transport.

labile sulphide: A compound from which H_2S evolves by treatment with acid, e.g. the sulphur of Fe-S proteins.

lability: Instability to heat, shear or other physical or chemical stress.

lac operon: The structural and associated regulatory genes that control the ability of *E. coli* to live on lactose; a subject of classical experimentation.

ladder: The pattern of bands seen on a polynucleotide sequencing gel that differ by integral numbers of nucleotides. (*see also* Maxam-Gilbert method (chemical cleavage method); Sanger method (dideoxynucleotide sequencing))

Laemmli gel electrophoresis: It stands for sodium dodecyl sulphate/ polyacrylamide-gel electrophoresis (SDS/PAGE))

laevo (levo)-rotation: (*see* optical rotation)

lagging strand: The discontinuous DNA strand that is synthesized at a fork during replication, the direction of synthesis of which is opposite to that of the movement of the fork. (*see also* leading strand; Okazaki fragment; semi-discontinuous)

LAGLI-DADG sequence: A conserved sequence in yeast mitochondrial unassigned reading frames, about 1kb long, that is part of an intron the 5'-end of which has the consensus sequence coding for the LAGLI pentapeptide and the 3'-end that for a DADG tetrapeptide.

lambda phage: (*see* bacteriophage)

lamellipodium (ruffled edge): A fin-like projection of the plasma membrane of a eukaryotic cell.

lampbrush chromosome: A chromosome in which the chromatin extends out from a central axis into many loops, where transcription and other activity occurs.

lap-joint: The non-covalent association of sticky-ended DNA duplexes.

large nuclear ribonucleoprotein particle (lnRNP): (*see* pre-mRNA splicing)

lariat: (*see* self-splicing)

laser Doppler flowmetry (LDF): (*see* near infrared spectroscopy)

laser-assisted desorption mass spectrometry: (*see* mass spectrometry)

last universal common ancestor (Luca): (*see* Woesean tree)

late stage: (*see* immediate-early stage)

latent: Hidden or cryptic, e.g. the enzymic properties that are expressed *in vitro* only when the vesicle that encapsulates the enzymes is lysed.

lathyrism: A skeletal malformation due to dietary -aminopropionitrile (present in the seeds of sweet peas, *Lathyrus odoratus*) which interferes with the proper cross-linking of collagen.

Laue crystallography (time-resolved crystallography): The use of X-rays or neutrons from a synchroton to collect diffraction data in very short times, sometimes measured in milliseconds, for time-resolved crystallography of macromolecules, especially proteins, in order to observe internal motions associated with conformational changes and/or catalysis.

lawn: The thick coverage of an agar (or other gel medium) plate with bacteria; usually as a test system for a lysogenic bacteriophage. (*see also* plaque-forming unit (PFU))

layer line: A feature of an X-ray diffraction pattern of a fibre. For fibres that contain aligned helices, the reflections appear as a pattern of spots arrayed in lines perpendicular to the fibre axis and the separation of the lines is proportional to the rise of the helix.

LD50: The dosage of a toxic material that is lethal to 50% of the animals, or cells, studied.

LD-MS: Laser-assisted desorption mass spectrometry. (*see* mass spectrometry (MS))

leader sequence: Signal sequence; the N-terminal portion of a secretory or membrane protein that assists it across the membrane of the rough endoplasmic reticulum, where it is synthesized, but which is cleaved from the protein even before the synthesis of the protein is complete. Mitochondrial and chloroplast proteins that are synthesized on ribosomes also have signal sequences that target them to their organelle; these may be N-terminal or internal, and may or may not be removed by proteolysis.

leading strand: The continuous DNA strand synthesized at a fork during DNA replication. (*see also* lagging strand; semi-discontinuous)

leaflet: One half of a lipid bilayer.

lectin: A protein, other than one of the immunoglobulins, that binds to the non-reducing ends of a specific oligosaccharide, e.g. that of a glycoprotein or cell surface. Lectins are usually of plant origin, although some have been identified in animal tissues and bacteria.

left-handed helix: (*see* right-handed helix)

leghaemoglobin: A single-chain oxygen-binding haem protein of legumes.

Leloir pathway: The metabolic route by which galactosyl and glucosyl units are interconverted; the transformation of galactose and ATP to glucose-1-phosphate via the galactokinase, galactose-1-phosphate uridyltransferase and UDP-galactose-4-epimerase reactions.

Lepore haemoglobin: A class of haemoglobin in which there has been a faulty crossing-over between the adjacent genes for - and -globins, producing a globin with the N-terminus of -globin and the C-terminus of -globin. *Anti-Lepore haemoglobins* are mutants with the N-terminus of -globin and the C-terminus of -globin.

leucine scissors: A supersecondary structure of some DNA-binding proteins; a coiled coil of two parallel -helices. Each is of about 30 residues in length and contains leucine residues at every seventh position, which places them all to the same side of the helix. The side chains interact with non-polar side chains of the other helix to hold the two together by hydrophobic forces.

leucine zipper: (= leucine scissors)

leuko-: A prefix derived from the Greek for white, signifying absence of colour.

leukotriene: One of a series of arachidonic acid metabolites; contains three conjugated double bonds and has been oxygenated through the action of a lipoxygenase.

levo-rotation: (= laevo (levo)-rotation; *see* optical rotation)

library: (*see* cDNA library; combinatorial; genomic library; shape library; small molecule library)

licencing: The limiting of replication of the eukaryotic genome to once per cell cycle. The proteins that impose this effect are termed *replication licencing factors* (*RLFs*), but their mode of action is unknown.

LIC-PCR: It stands for ligation-independent cloning of PCR products.

ligand: A small molecule or ion that binds to a protein or other structure.

ligand-mediated PCR (LMPCR): A method for amplification of a DNA fragment or a series of fragments with a single known primer-binding site. The fragment is denatured and hybridized with the primer. Extension of the primer with a polymerase that leaves a blunt end allows subsequent ligation to a *linker-primer*, which is a duplex oligodeoxynucleotide that will become the attachment site of a second primer in subsequent PCR cycles. (*see also* anchored PCR; inverse PCR)

ligase: One of a class of enzymes that join two substrate molecules in energy- (usually ATP-) dependent reaction, e.g. an amino acyl-tRNA synthetase, a carboxylase; in molecular biology, an enzyme that attaches the 3'-end of one polynucleotide to the 5'-end of another. (*see also* synthetase)

ligase chain reaction (LCR): A technique for detection and amplification of target DNA sequences. Two oligodeoxynucleotides are synthesized which between them are complementary to the entire target sequence, one to the 5'-side and one to the 3'-side. If the target sequence is present in the DNA sample under examination, the oligonucleotides will bind to it at moderate temperature with their ends abutting in the centre, and a heat-stable ligase will join them into a complete polynucleotide. No ligation will occur if the target sequence is absent or if the match between synthetic oligonucleotides and target sequence is imperfect in the region where they abut. At an elevated temperature the new polynucleotide dissociates from the original DNA template, and upon cooling to moderate temperature it and the original DNA serve as templates for a second cycle of hybridization, ligation and thermal dissociation. At each cycle there is a doubling of the number of new complete polynucleotides, so after 20 cycles there can be a 10^6-fold amplification. The half-sequences may be constructed with markers on the non-ligating ends to assist isolation (e.g. biotin) or detection (e.g. a fluorophore). (*see also* polymerase chain reaction (PCR))

ligation-independent cloning of PCR products (LIC-PCR): A method for cloning complex mixtures of PCR products to achieve a library of recombinant clones. PCR-amplified sequences, e.g. those bracketed by *Alu* sequences, are generated with primers that include sequences that contain a relatively long (e.g. 12 nucleotides) sequence devoid of G residues; similarly, a vector is cleaved at a restriction site and PCR-amplified with two different primers (different to prevent premature cyclization) that lack C residues in their 3'-ends. The potential inserts and the vector are trimmed with T_4 DNA 3'-5' exonuclease in the presence of dGTP and dCTP respectively to generate overhanging remnants of the primers, which were designed to permit insertion of the PCR fragments into the vector. As the overhangs are relatively long and quite stable, ligation before transformation is unnecessary.

ligation-mediated PCR: A PCR technique, similar to anchored PCR, applied to double-stranded DNA molecules for which the sequence of only one end is known, e.g. in *in vivo* footprint analysis to detect sites protected by a DNA-binding protein from dimethyl sulphate reaction and thus subject to subsequent chemical cleavage; a specific oligonucleotide is ligated to the unknown end and amplification is accomplished by the PCR procedure using primers complementary to the known end and the newly ligated oligonucleotide.

light chain: (*see* heavy chain)

light reaction: (= Hill reaction)

light scattering: A phenomenon of solutions of macromolecules; used analytically to determine the molecular mass and shape of a macromolecule from the effect on the intensity of scattered light of the wavelength of light and the angle of its scatter. In *Rayleigh scattering* the interaction of a photon with the molecule is elastic, i.e. there is no energy transfer from photon to molecule. In *Raman scattering* the interaction is inelastic, i.e. there is energy exchange, usually to vibrational energy (infrared) of the molecule, and the scattered light is then of longer wavelength.

light subunit: The smaller of the two ribonucleoprotein complexes that make up a ribosome; more generally, the smallest of the subunits of any complex. (*see also* heavy subunit)

light-chain library: (*see* repertoire cloning)

lignin: A random polymer in woody plants formed by oxidation of shikimic acid metabolites to form carbon-carbon and carbon-oxygen cross-links.

limit dextrin: The partially digested polysaccharide that is left after phosphorylase removes the outer tiers of glycogen or amylopectin and exposes the 16 branch points.

limiting polarization: The polarization of a fluorophore when it is immobile. (*see also* Perrin plot)

line width: The frequency range over which energy absorbance occurs, or over which a spectroscopic peak is observed; frequently measured at half the peak height.

LINE: It stands for long interspersed repeat element.

Lineweaver-Burk plot: A linearized display of kinetic data showing dependence of enzyme activity on substrate concentration, i.e. reciprocal of rate against reciprocal of substrate concentration, therefore also termed a *double-reciprocal plot*. (*see also* Eadie-Hofstee plot)

link protein: A protein that holds a subunit of a proteoglycan to the central hyaluronic acid molecule. (*see also* core protein)

linkage: In genetics, the more-frequent-than-random occurrence of two traits together due to their proximity on the same chromosome. The likelihood of a recombination event separating the two traits decreases with their increasing proximity on the chromosome. (*see also* linkage disequilibrium)

linkage disequilibrium: In genetics, the more-frequent-than-random occurrence together of two genetic loci within an individual. This may be due to their proximity on the same chromosome, to selective pressure on their common occurrence, or to close proximity on a chromosome of one locus with a second which is selected to occur with a third. The likelihood that a recombinant event can separate the two traits decreases with their increasing proximity on the chromosome. Linkage analysis is used to locate an unknown gene for a particular phenotype, often an inherited disease, by finding a closely linked genetic marker whose location is known. Association analysis is similar; it compares the frequencies of the different alleles at a specific locus with a range of phenotypes. (*see also* linkage)

linkage group: A collection of genes that are inferred to be located together on a single chromosome because of the pattern of their inheritance.

linkage map: A description of a genome derived from the screening of many phenotypes to determine the frequency with which genes occur together, and can then be placed in appropriate proximity to one another.

linker-primer: (*see* ligand-mediated PCR (LMPCR))

linker-scanner mutation: A method to assess the importance of various regions of a protein to its function by introduction of a dipeptide sequence into random sites of the protein. A plasmid that contains a structural gene is partially cleaved by a restriction endonuclease to achieve a single cleavage per coding region. The double-stranded DNA linkers, which are two-anticodon sequences between restriction nuclease-specific ends, are then patched into the gaps left by the cleavages.

linking clone: A tool for the localization of restriction fragments within a physical map of the genome of an organism. A cloned fragment that contains very few of an already rare restriction site is used as a hybridization probe for chromosomal DNA digested with the rare cutter.

linking number (L): A topological property of double-stranded DNA; the number of right-handed turns one DNA strand makes around the other. For relaxed DNA the number is equivalent to what it would be for linear, unstressed, DNA, i.e. 1 per 10 base pairs. L is equal to the twisting number plus the writhing number (L=T+W).

lipase: (*see* esterase)

lipid: A natural substance that is poorly soluble in water but is soluble in organic solvents; lipids include fatty acids, triacylglycerols, phospholipids, waxes and some hormones and vitamins.

lipid bilayer (bimolecular sheet): A synthetic or natural membrane in which amphipathic lipids are arranged in two layers with their non-polar chains directed inwards, towards each other, and their polar groups directed outwards, towards the aqueous phase. (*see also* fluid mosaic model (Singer-Nicolson model))

lipid peroxidation: The oxidation of polyunsaturated fatty acid side chains, initiated by a free radical such as the hydroxyl radical ($OH^{\bullet}$), to form a relatively stable carbon free radical, which reacts further with molecular oxygen to form a peroxy radical, and further yet with another fatty acid side chain to generate another carbon free radical. Such peroxidation causes rancidity in foods, damages membranes and, when it takes place in plasma low-density lipoprotein, can lead to atherosclerosis. (*see also* oxidative stress)

lipo-: A prefix derived from the Greek for fat.

lipocalin: One of a family of homologous proteins that bind lipophilic substances, including retinol-binding protein, -lactoglobulin, orosomucoid.

lipoma: A cancer of adipose tissue.

lipopolysaccharide (LPS): A constituent of the outer membrane of Gram-negative bacteria; composed of an outward-directed and highly variable oligosaccharide (the *O-antigen*), which is responsible for the antigenicity of the product, linked to a relatively invariable core oligosaccharide, in turn linked to a less polar moiety (*lipid A*) that is embedded in the membrane and is responsible for the endotoxicity and pyrogenicity of the product. *R strains* produce colonies with a rough surface and have truncated lipopolysaccharides; *S strains* produce colonies with a smooth appearance.

lipoprotein: A complex of lipids and apolipoproteins that is a transport form of lipids in blood. Lipoproteins are characterized by their density, which is determined by the lipid portion, and include high-, low- and very-low-density lipoproteins.

liposome: A synthetic micelle formed by dispersion of a polar lipid in aqueous solution in such a way that it forms into lipid-bilayer-encapsulated vesicles; useful for packaging of drugs or enzymes for introduction into cells.

LMPCR: It stands for ligand-mediated PCR.

location measure: An indication of the position of an amino acid residue within a proposed folding of a globular protein; the number of -carbons within a 14Å (1.4nm) radius, which varies from about 10 for a residue on the surface of a protein to about 80 for a residue well inside a large globular protein. (*see also* contact number)

lock and key model: A model for enzyme action that explains the basis of specificity as the exact fit of substrate to a site on the enzyme that is complementary in shape and electronic charge. (*see also* induced-fit theory (rack model))

locus control region (LCR): (*see* silencer)

locus: The position on a chromosome of a particular allele.

LOH: It stands for loss of heterozygosity.

London dispersion force: The weak interaction of neutral atoms due to the transient asymmetrical charge distribution within each atom. (*see also* van der Waals bond)

long interspersed elements (LINES): Retrotransposons, double stranded DNA sequences, 6-7 kb long, that occur in families of thousands of nearly identical copies which have apparently been randomly inserted into the mammalian genome. Short interspersed elements (SINES) are about 300 bp long. (*see* Alu sequence)

long terminal repeat (LTR): A polynucleotide sequence found at each end of an integrated retrovirus genome that contains the signals for expression of the viral genome.

long-chain fatty acid: (*see* medium-chain fatty acid)

loop: A packing structure of eukaryotic DNA that may be identical to a replicon. It is seen when the DNA is unfolded and visualized by electron microscopy, and shows the polynucleotide extending from closely spaced points of attachment to the nuclear matrix, which are presumed to be the terminators of replication. Also, in protein chemistry, a short polypeptide sequence of a protein that connects one region of secondary structure (-helix or¯-sheet) to another.

loss of heterozygosity (LOH): A test for chromosomal changes in tumour DNA and the theory that some tumours arise from inactivation of both alleles for an anti-oncogene. Heterozygotes, who already have one mutation, are most vulnerable. The tumour (e.g. breast cancer) genome, (presumably homozygous) is compared with the constitutional (e.g. leucocyte) genome (presumably heterozygous) for changes that can be interpreted as a change from the heterozygous to a homozygous state. With an appropriate probe for a site near a suspected mutation, restriction fragment length polymorphism analysis is used to detect the change.

loss-of-function: Descriptive of a experimental biological model that is designed to be unable to express a particular trait; e.g. a cell line treated with an antisense oligodeoxynucleotide. This contrasts with a gain-of-function model, which has an additional or supplemented activity; e.g. a transgenic animal.

low-angle scattering: A technique for determination of the size of particles in solution. The intensity of scattered X-rays from irradiation of a particle, e.g. a ribosome or a virus, is related to its size. In a variant technique, the solution of particles is irradiated with neutrons, the scattering of which varies with the mass of an atom's nucleus and thus can indicate the size of a phosphorus (i.e. nucleic acid)-containing particle or a ^{2}H-labelled protein.

low-barrier hydrogen bond: (*see* hydrogen bond)

Lowry protein assay: A refinement of the colorimetric method for protein determination that uses the Folin-Ciocalteu phosphomolybdotungstate reagent.

LPS: It stands for lipopolysaccharide.

LTR: It stands for long terminal repeat

luciferin: A potentially chemiluminescent substrate; the enzymic reaction of an enzyme (a *luciferase*) converts it to an excited state that decays with emission of visible radiation; e.g. the benzothiazole of fireflies which upon adenylation is decarboxylated and deadenylated with the emission of yellow light. The energy of photon emission derives from an oxidation with molecular oxygen that may be the enzymic or a non-enzymic reaction of the product of the enzymic reaction. There is no structural homology within the luciferins or luciferases, which have a wide biological distribution.

luminescence: Emission of a photon, of the same or lower energy than the energy that excited it, by an excited state of a chemical compound. (*see also* fluorescence; phosphorescence)

lyase: One of a class of enzymes that add one substrate across the double bond of another substrate, e.g. a decarboxylase, a dehydratase, an aldolase. (*see also* synthase)

lymphokine: A protein secreted by T-cells that functions in cell-mediated immunity to activate or inhibit leucocytes.

lymphoma: A cancer of lymphatic tissue.

lyo-: A prefix derived from the Greek for loose. (*see* lyso-)

lyotropic series (Hofmeister series): The arrangement of anions and cations according to their ability to modify the properties of other solutes such as proteins, probably via their effectiveness in altering the internal structure of the bulk water phase. *Chaotropes* are agents that disrupt water structure, and *cosmotropes* are agents that stabilize it.

lysate: The product of lysis; a broken-cell preparation.

lysis: The rupture of the membrane of a cell or bacterium, with the consequent loss of its constituents to the fluid in which it is suspended. (*see also* haemolysis (hemolysis))

lyso-: A prefix derived from the Greek for loose which indicates breakdown and dissolution.

lysogenic bacterium: A bacterium that hosts a prophage, which may lie dormant even for many generations but eventually may cause lysis.

lysosomal storage disease: A hereditary deficiency of one of the lysosomal enzymes involved in turnover of cellular components. An inability to degrade a substance often leads to the engorgement of the lysosomes with the undegraded material.

lysosome: A vesicle that contains many hydrolytic enzymes that are active in the characteristically low-pH environment.

lytic phage: (*see* bacteriophage)

M

-mer: A suffix derived from the Greek for part that indicates degree of polymerization; e.g. tetromer, 40-mer. (*see* -mere)

-mere: A suffix derived from the Greek for part which indicates a subcellular structure. (*see* -mer)

-meter: A suffix derived from the Greek for measure.

-morph: A suffix with the same derivation as morpho-.

-mycin: A suffix with the same derivation as myco-.

M period: (*see* cell cycle)

mAb: (= monoclonal antibody (mAb))

macro-: A prefix derived from the Greek for large; e.g. macrophage, macroenzyme

macroenzyme: A normal, active enzyme that appears in serum as a conjugate, often with a specific autoantibody, thus giving it a greatly increased molecular weight.

macrolide: A moiety of some antibiotics (e.g. erythromycin, methymycin) which consists of a large lactone ring, synthesized largely or entirely from propionate-derived units. (*see also* polyketide; propionate rule)

macromolecule: A compound or complex, usually a polymer such as a protein, nucleic acid or polysaccharide, or a covalent or non-covalent complex of any of these.

macrophage: One kind of phagocytic cell. (*see* innate immunity; opsonin)

MAD: (= multi-wavelength anomalous diffraction (MAD))

magainin: One of a group of anti-microbial oligopeptides secreted by the skin of frogs. (*see also* cecropin)

magic spot nucleotides: Guanine nucleotide derivatives (ppGpp and pppGpp) that mediate the cessation of production of rRNA in the absence of an essential amino acid, a phenomenon present in *stringent bacteria* but absent from *relaxed bacteria.* (*see also* alarmone)

magnet-assisted subtractive technique (MAST): A method for detection of the tissue specificity of gene expression. Two tissues are compared: a deficient tissue from which single-stranded cDNA (driver cDNA) is prepared and attached to magnetic beads to which are also attached poly(T) sequences, and a normal tissue from which cDNA (tracer cDNA) is prepared. An excess of the driver cDNA hybridizes with the tracer cDNA preparation from all except those mRNAs in which the first tissue is deficient, and will be removed by magnetic separation of the beads. The resulting cDNA can be used to create a subtraction library for examination of tissue-dependent expression. (*see also* subtractive DNA cloning)

magnetic circular dichroism (MCD): A technique to characterize the electronic environment of a metal ion, especially in a metalloprotein. The circular dichroism (CD) spectrum of a sample is recorded in a magnetic field parallel to the light path. Differences in the CD spectrum due to the magnetic field are interpreted as being due to Zeeman splitting of the excited state (if temperature-independent) or of the ground state, in which case much greater detail is available at very low temperatures where only one ground state is populated.

Maillard reaction: (= browning reaction (Maillard reaction))

maintenance: The process by which some genes are rendered non-equivalent. The paternal or maternal allele is not expressed (allelic exclusion), or is expressed differently in different tissues. The phenomenon is possibly directed by an imprinting box, a parent-specific polynucleotide sequence that instructs an imprinting factor, which may act by methlyation of differentially methylated regions (DMRs), and which will reversibly activate or inactivate the gene inherited from only one parent. Imprinting passes through erasure of methylation of both parental and maternal chromosomes during germ cell development, establishment of methylation, according to the sex of the gamete, and after fertilization, maintenance of the methylation status of the genome during the mitoses of embryogenesis.

major groove: The wider of the two helical spaces on the surface of an A- or B-DNA double helix. The other helical space is the *minor groove*.

major histocompatibility complex (MHC): The products of genes grouped together on a chromosome that determine whether transplanted tissues will be accepted (if from an individual with the same MHC) or rejected. In the human, they are known as *human leucocyte antigen molecules* (*HLAs*). MHC gene products are highly variable intrinsic membrane proteins of the immunoglobulin superfamily and consist of two types: *class I* form non-covalent heterodimers with $_2$-microglobulin and, along with a foreign protein such as a viral antigen, comprise a recognition complex for a cytotoxic T-lymphocyte; *class II* consist of dimers of highly variable such proteins, and occur in macrophages and B-lymphocytes.

MALDI-MS: Matrix-assisted laser desorption/ionization mass spectrometry. (*see* mass spectrometry (MS))

malignant: Descriptive of a tumour that has a high tendency to become progressively larger and more invasive and to undergo metastasis.

mammalian artificial chromosome (MAC): *See* human artificial chromosome (HAC)

manganese centre: In the photosynthetic photosystem II, a complex of four manganese atoms separated by oxygen atoms that accumulates a positive charge that is used to oxidize water to molecular oxygen. (*see also* charge accumulator model)

manometry: A largely obsolete technique for measurement of respiration of tissue slices or other biochemical preparations. The volume of oxygen taken up is measured in a *Warburg apparatus*, a closed system that consists of a flask with a central well in which NaOH is placed to absorb carbon dioxide. In the absence of alkali, the volume of evolved carbon dioxide is offset by oxygen taken up and is evaluated by difference.

Manton-Gaulin homogenizer: An apparatus for disruption of cells, especially the large quantity of bacteria or yeast that result from a fermentation, by pumping a paste of the collected cells through an orifice of adjustable size in the head of a reciprocating piston so that the shearing forces that develop break apart the cells.

MAP kinase: Mitogen-activated protein kinase; also known as extracellular-signal-regulated protein kinase (ERK).

MAP: (= methyl-accepting protein (MAP); microtubule-associated protein (MAP))

mapping: The creation of an outline of locations of genetic markers (genes or other polynucleotide sequences) within the structures of the chromosomes. The methodology determines the resolution of the map. Being dependent upon visual markers such as break points in the chromosomes, *cytogenetic mapping* is the least well resolved. *Genetic linkage mapping* is dependent upon linkage of markers, i.e. the frequency with which markers appear together within a population. *Somatic cell hybrid mapping* is dependent upon cell lines that contain a known whole human chromosome in a rodent cell or, in the case of radiation reduction hybrids, fragments of a known human chromosome interspersed within the chromosomes of a rodent cell.

marker rescue: A technique to localize the site of a mutation within a gene. A polynucleotide, the 'marker', from the known wild-type sequence is allowed to recombine with the mutant genome such that, if it includes the site of the mutation, it 'rescues' the mutant by causing it to revert.

Masbauer spectroscopy: A technique for studying the chemical environment of some metal atoms in solid samples, especially ^{57}Fe. The metal is irradiated with soft -radiation and emits lower-energy -radiation that is indicative of its valence state, electronic and magnetic environment.

mass spectrographic immunoassay (MSIA): A technique for characterization of antigens, especially when more than one may bind to an antibody. After isolation of an antigen from a solution by using an immobilized antibody, the adsorbed antigen is eluted and analysed by mass spectrometry

massively parallel signature sequencing (MPSS): A method for automated identification by sequence and quantitative evaluation of expressed genes, which is unaffected by sequence-dependent variations in hybridization efficiencies. A microbead cDNA library, one DNA fragment per bead, is immobilized in a planar flow cell that allows exposure to a sequence of reagents and scanning of the cell to record the positions of attached fluorescent tags. The entire library is sequenced simultaneously in cycles; during the cycle, each bead reports a four-base sequence, so four or five cycles are sufficient to unambiguously identify a fragment. To initiate the procedure, an endonuclease exposes four-base overhangs. The sequences of these overhangs are identified in four steps to identify the first, second, third and fourth base by ligation to a characteristically tagged adaptor oligonucleotide. In the first step, 16 adaptors that are presented have the first (5') nucleotide paired with a 3' sequence that is designed to accept one of four fluorescent tags; the next three 5' nucleotides are randomized to allow all possible template sequences to bind one of the set of 16 oligonucleotides. The positions of the fluorophores within the flow cell are recorded and the adaptors cleaved. In the next three steps, the second, third and fourth positions are similarly determined. The cycle ends with the exposed 5' ends being shortened to allow the next four-base sequence to be determined in the next cycle. Because of the numbers of beads that are accommodated by this method, each cDNA may be represented up to hundreds of times; the number of beads that contain each unique sequence, therefore, is a measure of gene expression.

mast cell: A cell that contains histamine-filled secretory vesicles.

mating type: In effect, a sex of a haploid yeast cell that restricts it to forming a diploid cell only with a haploid of a different mating type.

mating-type switching: A phenomenon in the behaviour of yeast and other organisms that rapidly change from the haploid to the diploid state. In yeast the two mating types are a and ; when two a or two cells mate, the diploid form results in daughter cells of mating type opposite to that of the parent cells.

matrix metalloproteinase (MMP): One of the family of Zn-containing proteins that includes fibroblast and neutrophil collagenases, gelatinase and stromelysin. (*see also* cysteine switch)

matrix-assisted laser desorption/ionization mass spectrometry: (*see* mass spectrometry)

matrix-attachment region: (*see* AT queue)

matrixin: (= matrix metalloproteinase)

maturase: A nuclease involved in RNA processing; it is formed by translation of a fungal mitochondrial mRNA that is produced after a first intron is excised leaving an mRNA in which the open reading frame continues through a second intron and codes for the maturase. The enzyme, by participating in the cleaving out of the second intron, destroys its own mRNA.

Maxam-Gilbert method (chemical cleavage method): A technique for sequence analysis of DNA in which four chemical reactions are applied separately, each to cleave the polynucleotide randomly at one of the four bases. Subsequent polyacrylamide-gel electrophoresis separates the products according to chain length, and reveals the distance from the radiolabelled 3'- or 5'-end to the chemically modified base. (*see also* Sanger method)

maximum velocity (Vmax): The limiting rate for an enzymic reaction, shown when it is saturated with substrate. (*see also* Michaelis-Menten kinetics)

MCD: (= magnetic circular dichroism (MCD))

MCP: Multicatalytic proteinase. (= proteasome)

MDR: (= multi-drug resistance (MDR))

mechanism: A flexible term applied to the details of a process that links initial and final states, e.g. the mechanism of respiration, the mechanism of fatty acid oxidation, the mechanism of -oxoacyl-coenzyme A dehydrogenase activity.

mechanism-based inhibitor: An enzyme inhibitor that differs from the classical substrate or product analogue. One category comprises *transition-state analogues* that do not form covalent links to the enzymes but bind very tightly to them; the other category contains *enzyme-activated*, k_{cat} or *suicide inhibitors*, which require the action of the enzyme upon them to convert them into alkylating or other reactive agents that can form covalent complexes with the enzyme.

medial Golgi: (*see* Golgi apparatus)

mediated transport: (= facilitated diffusion)

mediator: A complex of proteins that allows initiation of cell- and/or cell cycle-specific control of eukaryotic transcription. The complex bridges a (usually) upstream promoter DNA sequence bound to an enhancer or silencer, and RNA polymerase II at the initiation site. (*see* origin recognition complex)

medium: The solid or liquid substratum on which, or in which, cells or organ explants can be made to grow; a medium may include well defined factors such as salts, amino acids and sugars, as well as less well defined factors such as serum or blood.

medium-chain fatty acid: A fatty acid with a skeleton of 4-12 carbon atoms. Those with fewer carbon atoms are *short-chain fatty acids*; those with more (up to 20 carbon atoms) are *long-chain fatty acids*; and those with skeletons of 20 carbon atoms or more are *very-long-chain fatty acids*.

medRNA: Mini-exon-derived RNA (*see* trans-splicing)

medulla: The inner core of an organ, e.g. adrenal medulla. (*see also* cortex)

mega-: A prefix (abbreviated M) which denotes 10^6; e.g. megabase pair, 1 Mbp, 10^6 bp.

megadose: An amount of a dietary supplement, such as a vitamin, mineral or essential amino acid, that is orders of magnitude greater than the officially recommended dosage.

megakaryocyte: A large cell of bone marrow characterized by a multi-lobed nucleus, usually with eight times the haploid number of chromosomes; the precursor of a platelet.

meiosis: The process of formation of haploid gametes from a diploid cell.

melano-: A prefix derived from the Greek for black.

melanocyte: An epithelial cell that contains melanin pigment granules, *melanosomes*.

melanosome: (*see* melanocyte)

melting temperature: In nucleic acid and protein chemistry, and by analogy with organic chemistry, the temperature that is the mid-point for the thermal unfolding of a nucleic acid or protein; in the case of nucleic acids, usually measured by the hyperchromic effect.

membrane asymmetry: A phenomenon of membrane proteins and gangliosides involved in signal transduction and transport processes; as they are oriented towards the inside or outside, their asymmetry is appropriate to their function.

membrane protein secretase: A sheddase, a matrix metalloprotease that cleaves an exposed non-globular region (the stalk) of a protein attached to the outer surface of the plasma membrane, often a transmembrane protein, to release a large fragment of it, an ectodomain, into the extracellular space.

membrane skeleton: The matrix of proteins and fibres that underlies the lipid bilayer of a plasma membrane; it contains actin filaments, actin-binding proteins and/or cell-specific structural proteins.

membrane-associated thermosynthesis (MAS): (*see* thermogenesis model)

memory B-cell: A long-lived activated B-cell that does not differentiate into a plasma cell until stimulated further.

merocrine secretion: (= exocytosis)

mesenchyme: The embryonic precursor of connective tissue.

meso-: A prefix derived from the Greek for middle which indicates something intermediary.

meso-carbon: A prochiral centre; a tetrahedral carbon to which are attached two like and two other unlike groups. (*see also* Ogston hypothesis)

mesoderm: One of the three primordial germ layers formed during early embryogenesis; a precursor of muscle, adipose tissue, blood vessels, the gastrointestinal tract, etc. (*see also* ectoderm; endoderm)

mesophile: (*see* extremophile)

messenger RNA: (= mRNA)

meta-: A prefix derived from the Greek for with or after, which indicates association or intermediacy. In benzene derivatives, meta- signifies the relation of substances on nuclear carbon atoms two removed from each other. In inorganic chemistry, meta- signifies a compound obtained by dehydration, e.g. meta-phosphate (PO_3-) from ortho-phosphate ($PO4^{-3}$).

metabolic channelling: (*see* metabolon)

metabolic pathway: A sequence of metabolic reactions that transforms a substrate, e.g. glycolysis, -oxidation, gluconeogenesis.

metabolism: The sequences by which foodstuffs are degraded for the energy that is released and for transformation into cellular components.

metabolite: An intermediate or end product of intermediary metabolism.

metabolon: A proposed multienzyme complex that is responsible for metabolite channelling, so as to eliminate or minimize loss of efficiency due to the otherwise necessary diffusion of substrates to and of products from the enzymes of a metabolic pathway.

metabomics: (*see* omic research)

metabotropic: (*see* ionotropic)

metal-binding finger: (= zinc finger (metal-binding finger))

metalloproteinase: A type of peptidase that has a metal ion at its active site. (*see also* aspartate proteinase; cysteine proteinase (thiol proteinase); serine proteinase)

metastasis: The shedding of malignant cells from a tumour, which permits them to establish new colonies in other tissues.

methionine bristles: The cluster of methionine side chains that are proposed to protrude from the surface of a signal recognition particle and provide a plastic hydrophobic environment that can interact with the hydrophobic residues of the signal sequence of a nascent protein.

methyl trap: The accumulation of 5-methyltetrahydrofolate in vitamin B_{12} deficiency or related conditions (e.g. pernicious anaemia) due to the inability to transfer the methyl group to homocysteine by the B_{12} coenzyme-dependent transferase.

methyl-accepting protein (MAP): One of a group of reversibly methylated proteins of some bacteria that are involved in the chemotactic response to chemoattractants.

methylation analysis: Any technique to detect sites at which bases of DNA are methylated, e.g. by analysis of DNA digested by isoschizomeric pairs of restriction endonucleases.

methylation-specific PCR (M-PCR): A technique for mapping methylation patterns of CpG islands of genomic DNA. Genomic DNA is first treated with sodium bisulphite ($NaHSO_3$) to convert cystosine residues, but not 5-methylcytosine residues, into uridines. Primers specific for the untreated sequence will amplify the methylated sequence for detection, while pirmers for the treated sequence will amplify the unmethylated sequence.

methyl-protection footprinting: (*see* footprinting)

metro-: A prefix with the same derivation as -meter.

MHC: (= major histocompatibility complex (MHC))

micelle: A very small aggregate of matter that is dispersed in solution; often of lipoid material stabilized by detergents, or even of detergent only.

microarray separation: A method for sorting uniformly charged macromolecules, particularly nucleic acids, by size. Unlike conventional electrophoretic separations, the method uses a gel phase. By bonding a flat silicon surface to another that has been microlithographed with a pattern of raised rectangles or asymmetric diamonds, an array of sub-micron channels is created. As the mixture of macromolecules enters the array at a top corner of the plate, or chip, an electric field pulls the molecules in the vertical dimension, but the pattern of channels is such that the vertical flow repeatedly encounters the rectangle or diamond obstacles and is split vertically and horizontally away from the site of injection. Larger molecules diffuse less and consequently flow in the direction of the electric field, while smaller molecules diffuse faster and are disproportionately directed horizontally. As the solution emerges from the chip, the smaller macromolecules emerge furthest from the corner at which the solution had been applied.

microbody: (= peroxisome (microbody))

microbore HPLC: (*see* high-pressure (high-performance) liquid chromatography (HPLC))

microcluster: An aggregate of several (2-10) membrane receptors that forms in response to cell stimulation; such an aggregate is a transitory state between the receptors being dispersed in the membrane and their formation into much larger aggregates, patches that will undergo endocytosis.

microheterogeneous: (*see* monodisperse)

microhomology: (*see* double-strand-break repair model)

microreversibility: (= microscopic reversibility)

microsatellite: One of many short, highly polymorphic, non-coding sequences, e.g. poly(TG), found well spaced throughout the genome that can serve as landmarks during physical mapping.

microscopic reversibility: A principle from molecular physics that holds that each proposed step of a catalytic cycle must in theory be reversible, and particularly that the forward and reverse reactions must pass through the same transition state.

microsome: An artifactual vesicle formed from the endoplasmic reticulum when cells are disrupted.

microtitre plate: A transparent tray with 96 or 384 wells in which chromogenic reactions may be run and then automatically scanned to record changes in absorbance.

microtubule: A large (30nm diameter) rigid component of the cytoskeleton that is built of - and -tubulin subunits and associated proteins and serves as a component of cilia, eukaryotic flagella and mitotic spindles.

microtubule-associated protein (MAP): One of the non-tubulin proteins that are associated with microtubules in some of their cellular sites.

microtubule-organizing centre (MTOC): A cellular structure that serves as a site for nucleation of microtubule growth. (*see also* **centriole**)

microvillus: A finger-like projection from the apical membrane of some epithelial cells, especially those of the intestine.

mimetic: A non-peptide synthetic molecule that reproduces some features of a natural peptide, e.g. it may bind strongly to an antibody by placement of the same functional groups, in the same relation to each other, as are found in the peptide antigen.

mimotope: A synthetic peptide that binds to an antibody against an unidentified antigen; usually a peptide selected from a large library of peptides. (*see also* random peptide library)

mineralocorticoid: An adrenal steroid that promotes resorption of Na^+ by the kidney tubule; notably, aldosterone.

miniantibody: (= single-domain antibody (dAB))

minichromosome: A circular double-stranded DNA structure packaged with protein into nucleosomes or higher-level organization of chromatin. (*see also* double minute chromosome (DMC))

minimal molecular mass: A value for the molecular mass of a macromolecule calculated from its content of a metal ion, end group or other feature present in an unknown stoichiometric ratio, e.g. the minimal molecular mass of haemoglobin calculated from its iron content is: 55.84g/g-atom of iron 100/0.34% iron in haemoglobin=16400g/mol.

minimum daily requirement: The quantity of nutrients required each day in the diet by average individuals. (*see also* recommended daily allowance (RDA))

minisatellite variant repeat-specific priming PCR (MVR-PCR): A method for characterization of DNA samples of individuals for forensic purposes, based on sequence variations within the repeating units of variable number tandem repeats (VNTRs). PCR primers are designed for a common flanking region and each of the variant repeating units; the latter primers are initially used in low enough concentration so that on average only one binds per template; thus upon amplification and electrophoresis, a ladder is produced for each primer pair that indicates the position of the variant along the length of the VNTR. The sequence of the variants derived from ladders for all primer pairs is subject to great diversity, and is both characteristic of an individual and capable of digitalization for data storage, search and retrieval.

minisequencing: A multiplex method for scanning genomic DNA samples for the presence of specific mutations. PCR primers complementary to the mutant and wild-type sequences are immobilized at coded loci on a surface and exposed to genomic DNA samples. Incorporation of a fluorescent marker during PCR amplification will indicate by the presence of fluorescence at a locus, the existence there of the mutant or wild-type template in the genomic DNA to which it had been exposed.

minor groove: (*see* major groove)

minRNA: (*see* trans-splicing)

minus-strand: (*see* plus-strand)

(-)-strand: Minus-strand (*see* plus-strand)

mismatch analysis: (= heteroduplex analysis (mismatch analysis))

mismatch repair analysis: An assay system for detection of single-base mismatches in a DNA fragment of up to 10 Kb in length, which depends upon an *Escherichia coli* repair system that detects mismatched bases between a newly synthesized (unmethylated) DNA strand and an unreplicated (methylated) strand. The unmethylated DNA strand in a hybrid of the two is nicked, digested and resynthesized on the methylated template. This system, however, does not replace insertions or loops larger than 5 bases. A mutated DNA fragment is inserted into a plasmid that contains a reporter gene, such as *LacZ*, which can be used to identify colonies that express the plasmid, and is then transformed into an *E. coli dam-* strain, which results in an unmethylated plasmid.

The wild-type DNA fragment is inserted into the same plasmid in which a 5-base sequence interrupts the reporter gene; this is transformed into a *dam+* strain, which fully methylates the plasmid. The two strains are grown, the plasmids isolated, mixed, denatured and reannealed. Fully methylated homoduplexes are digested with *Dpn*I and unmethylated homoduplexes are digested with *Mbo*I. Surviving heteroduplexes, which contain both the intact and the interrupted reporter genes, are circularized, ligated and transformed into *E. coli*. If there is no mismatch, both strands of the plasmid will be replicated and the intact reporter gene will be expressed. However, if there is a mismatch, the non-methylated strand, along with the intact reporter gene will be digested and will not be expressed.

mismatched PCR: A method for detection of a point mutation that neither produces nor destroys a restriction site; intended especially for screening large numbers of individuals. One of the PCR primers is designed so that it is complementary to the sequence immediately upstream (or downstream) from the mutation, and near its 5'-end (or 3'-end) contains a mismatched base. The mismatch and the nearby mutated base (but not the normal base) when amplified generate a new restriction site. The amplified mutant DNA (unlike the amplified normal DNA), when digested with appropriate endonucleases, electrophoresed and probed with an appropriate labelled oligonucleotide, will reveal a new, smaller restriction fragment. Alternatively, the mismatched primer may be designed to create a new restriction site with the normal, but not the mutant, DNA. (*see also* restriction-site-generating PCR (RG-PCR))

missense mutation: A mutation that substitutes one amino acid residue for another.

missense suppressor: A mutation that produces a tRNA that can recognize a missense mutation and insert into the protein sequence an amino acid residue that permits phenotypic expression of the gene.

mitochondrial permeability transition (MPT): The disruption of normal mitochondrial function by opening of non-specific MPT pores (MPTPs). This allows the free diffusion of small molecules and ions and if prolonged, can lead to necrosis. Another outcome may be the release of cytochrome *c* which by activating several caspases, can lead to apoptosis.

mitochondrial wave: A phased depolarization of the mitochondrial surface due to changes in Ca^{2+} permeability.

mitochondrion: An organelle of eukaryotic cells that is the site of oxidative phosphorylation, the tricarboxylic acid cycle, -oxidation, etc., and is composed of a relatively permeable outer membrane and a deeply indented inner membrane.

mitogen: A compound that induces cell division.

mitoplast: A preparation made from mitochondria in which the outer membrane is stripped away by detergents and the inner membrane with the enclosed matrix is left intact.

mitosis: Cell division that produces daughter cells with the same number of chromosomes as the parent cell.

mitosol: The soluble matrix that occupies the space inside the mitochondrial inner membrane.

mixed disulphide: An oxidation product of two dissimilar thiol compounds.

mixed inhibition: (= non-competitive inhibition (mixed inhibition))

mixed-function oxidase (mono-oxygenase): An enzyme that reduces molecular oxygen by incorporation of one atom into water and the other into another substrate; e.g. phenylalanine hydroxylase. (*see also* dioxygenase)

mixotrophic: In microbiology, descriptive of an organism that requires organic material and light for growth. (*see also* autotrophic; heterotrophic)

MMP: (= matrix metalloproteinase (MMP))

mobile barrier: A proposed mechanism of membrane transport. A transmembrane protein, the activated porter, exposes its binding site for the transported molecule to one membrane face and then to the other. In the case of a *uniport porter*, activation does not depend upon the solute, so transport can be effected in either direction; a *symport porter* is activated only when both (or neither) of the two solutes are bound; an *antiport porter* is activated when either, but not both, of the two solutes is bound. This proposal is often coupled with the concept of *solvation substitution*, the identification of solute binding as the replacement of bound water by specific liganding groups of the porter, e.g. chelation of ions, to help explain the specificity of the porter and how it maintains a barrier against transport of water across the membrane. (*see also* mobile carrier)

mobile carrier: A proposed mechanism of membrane transport similar to the mobile barrier mechanism, but in which the porter is imagined to be a revolving door that accepts one (uniport) or two (symport) solutes, or distinguishes between membrane faces (antiport).

mobile genetic element: A transposon or an insertion sequence; a polynucleotide sequence that can move from a chromosome or plasmid to another chromosome or plasmid. (*see also* intron homing; protein splicing)

mobility shift assay: (= gel shift assay (electrophoretic mobility shift assay; EMSA))

model substrate: A synthetic substrate that often contains only the minimum necessary features to satisfy the specificity of the enzyme, but sometimes incorporates chemical features to facilitate detection of enzyme activity; e.g. a *p*-nitrophenyl ester as an endopeptidase substrate, as the *p*-nitrophenolate product is intensely coloured.

module: In protein chemistry, a building block from which larger proteins are constructed; a recurring folding pattern in proteins, examples of which may have amino acid sequences that are homologous or non-homologous, e.g. the kringle, the EF hand. (*see also* motif)

moiety: A fragment of a molecule, especially one that comprises an identifiable unit, e.g. an acetyl or pyridoxal phosphate group, a regulatory subunit.

molar: The concentration of 1 mole per litre; 1 mole being Avogadro's number (6.02×10^{23}) of molecules, which is equivalent in mass to the sum of the atomic masses (in grams) of the constituent atoms; e.g. 1 M NaCl is a concentration of 58.45 g/l (Na = 22.99, Cl = 35.46). Sub-molar concentrations are: millimolar (mM) 10^{-3} mol/l; micromolar (M) 10^{-6}mol/l; nanomolar (nM) 10^{-9} mol/l; picomolar (pM) 10^{-12} mol/l; femtomolar (fM) 10^{-15} mol/l; attomolar (aM) 10^{-18} mol/l; zeptomolar (zM) 10^{-21} mol/l.

molar absorption coefficient (extinction coefficient): The (calculated) absorbance of a 1M solution of a chromophore present in a 1cm light path. (*see also* Beer-Lambert equation)

mole: (*see* molar)

mole percent excess: A quantitative measure of the concentration of a stable isotope, analogous to the specific radioactivity of a radioisotope; the enrichment of that isotope, as a percentage of all isotopes, over and above its usual occurrence in nature.

molecular accordion: A reporter molecule; a fluorophore-labelled linear polymer that interacts with natural and synthetic phospholipid membranes, and whose fluorophores respond to the polymer's physical environment. The polymer attaches to the membrane through its aromatic fluorophores, which are sparsely spaced along its length and allow the intervening polymer to loop out into the solvent. The fluorescence intensity is greatly enhanced by the mutual proximity of the fluorophores, so physical forces that promote close association of the polymer loops (e.g. a lowered temperature or a solvent that interacts poorly with the polymer) will close up the loops, draw the fluorophores closer together and increase fluorescence, while a strongly interacting solvent or a higher temperature will cause the loops to relax, separate the fluorophores, and decrease fluorescence.

molecular archaeology: (*see* jumping PCR)

molecular channel: A proposed feature of multienzyme complexes; the product of catalysis by the first enzyme of the metabolic sequence travels through the channel to the active site of the next enzyme, etc.

molecular chaperone: Enzymes and other cellular protein complexes that ensure the proper folding of nascent or improperly folded proteins. There are two classes of chaperone: the heat shock protein (hsp) 70 family and chaperonins. The hsp 70 family bind to peptides as they form on the ribosome, or immediately after synthesis, and prevent inappropriate folding. Subsequently, the polypeptide folds spontaneously, or with the aid of chaperonins. Non-steric chaperones act by reversibly binding to unfolded polypeptide chains and prevent aggregation; steric chaperones, e.g. peptidyl proline isomerases, act to direct correct folding.

molecular chronometer: A phylogenetic marker; a highly conserved protein (e.g. ubiquitin) or nucleic acid (e.g an rRNA) whose rate of mutation is constant, and which can therefore be used to construct phylogenetic trees.

molecular clamp: A protein, or part of one, which firmly holds a ligand by surrounding it by a complementary surface.

molecular clock hypothesis: The proposal, so far experimentally unverified, that mutations have occurred at random and at a constant frequency in all species throughout the billions of years of biochemical evolution. Successful mutations, i.e. those that persist, are only those that do not adversely affect stability or function.

molecular dynamics: The computer-simulated quantitative evaluation of a molecular model by assignment of force constants to bonded and non-bonded interactions between neighbouring atoms so that bond distances and angles are confined to an acceptable range and crowding is minimized; often coupled with *energy minimization*, which makes small adjustments in the locations of atoms to optimize these interactions.

molecular evolution: The changes in gene polynucleotide sequences and resulting protein structures that have paralleled the evolution of species.

molecular filter: The function of specificity in membrane protein trafficking that directs membrane components to their appropriate locus; a phenomenological, rather than a morphological, designation.

molecular imprinting: The formation of very specific binding, or even catalytic, characteristics of a matrix by polymerization of a suitable monomer in the presence of a ligand or a transition-state analogue. (*see also* bio-imprinting)

molecular matchmaker: A protein that effects complex-formation between another protein and DNA by catalysis of an energy-dependent conformational change in at least one of the pair; it thereby increases affinity and permits binding that would not occur spontaneously. (*see also* chaperone machine)

molecular mechanics: (= molecular dynamics)

molecular model: A physical representation, e.g. balls and sticks, used to represent the structure of a compound or macromolecule, or a computer graphics simulation of such a structure. More sophisticated versions test structure- function relationships by simulation of ligand binding and conformational changes (*see also* docking; molecular dynamics)

molecular modelling: Any of the computational approaches for relating chemical or physical properties to structure, e.g. design of agonists or antagonists from the X-ray crystallography data of a receptor protein.

molecular motor: A complex of proteins that is responsible for energy, usually ATP hydrolysis-dependent, intracellular movement of an organelle or smaller structure along a cytoskeletal filament; e.g. the microtubule-anchored spindle motor, the actomyosin engine. The action of a motor is conceived as a duty cycle of working and recovery strokes, like the action of oars in a rowboat. During the recovery stroke, a point of contact is broken ('the oars come out of the water') to prevent reversal. The duty ratio, then is the fraction of total cycle time that contact at any single point is maintained; it can vary between 0 and 1, and for a bundle of myosin filaments with many heads (like the many oars of a trireme, approaches 0.

molecular organelle: A very large, non-membrane-limited aggregate of subunits that in prokaryotes functions like an organelle. They occur also in eukaryotes, e.g. the proteasome, which as a 26 S multimer is a cylinder 40 m long and 20 m in diameter.

molecular pharming: The production of human proteins of pharmacological interest in the milk of transgenic farm animals.

molecular resonance force microscopy (MRFM): A technique similar to atomic force microscopy, which, in theory, can detect and map magnetic resonance in paramagnetic nuclei of macromolecules. The MRFM resonator probe, mounted on a flexible cantilever in proximity to a test molecule, detects resonance at a distance of up to 100 Å. The tip of the probe scans the molecule, which is anchored to a surface, and the magnetic field emanating from the tip evokes a resonance in the sample which is detected by a minute vibration of the cantilever. An Å-scale map of the resonance should provide a three-dimensional image of the test molecule.

molecular ruler: A method for measurement of distances on a molecular scale; the estimation of distances between a fluorescence energy donor and receptor, evaluated from the efficiency of resonance energy transfer, which decreases according to the sixth power of the distance that separates them. (*see also* resonance energy transfer)

molten globule intermediate: A stable intermediate between the denatured and native states of a protein which is characterized by a high content of secondary structure but poorly defined tertiary structure, hydrophobic residues exposed to solvent, compactness intermediate between the denatured and native states, high sensitivity to proteases and, if an enzyme, no activity. (*see also* framework model; funnel concept; hydrophobic collapse; two-state hypothesis)

molten globule: (*see* funnel concept)

mono-: A prefix that indicates one basic unit, without any polymerization, as in mononucleotide, monosaccharide. (*see also* oligo-; poly-)

monocistronic mRNA: An mRNA that codes for a single protein; characteristic of eukaryotes. (*see also* cistron; operon; polycistronic mRNA)

monoclonal antibody (mAb): An immunoglobulin preparation that is completely homogeneous, due to its formation by daughters of a single progenitor cell that has been programmed for the synthesis and secretion of one specific antibody. (*see also* polyclonal antibody)

monocyte: One type of non-granular leucocyte; a precursor of macrophages and osteoclasts.

monodisperse: Descriptive of a population of polymers in which all individual molecules have the same covalent structure and molecular size. It is contrasted with *polydisperse*, descriptive of a population of polymers that have the same repeating unit(s) and the same covalent structure, but vary in molecular size. *Paucidisperse* describes a population of related polymers that share the same basic covalent structure, and perhaps function, but show a low level of dispersity. *Microheterogeneous* describes a population of polymers with the same primary structure, and possibly the same molecular size, but with limited modifications of some monomeric units.

mono-oxygenase: (= mixed-function oxidase (mono-oxygenase))

monoterpene: (*see* terpene)

morgan: (*see* centimorgan)

morpho-: A prefix derived from the Greek for form, which indicates structure.

morphogen: A molecule, secreted by a tissue, that acts by a concentration-dependent mechanism to induce position-specific patterning of distant (non-adjacent) cells during early embryogenesis.

mosaic protein: A protein composed of discrete domains, each encoded by a different exon, or sometimes by several exons. (*see also* exon shuffling)

mosaicism: The expression of a mutation in some somatic cells but not others of an organ or tissue.

mother liquor: The solute-depleted, but still saturated, solution which remains after crystallization of a solute from the solution.

motif: A recurring pattern of protein folding, e.g. a homeobox, a zinc finger. A higher order of protein structure is a module which consists of several motifs and units of secondary structure; in particular cases, it may be equivalent to a domain, which in a globular protein is a region that has its own tertiary structure, is stable independently of the rest of the protein and is often connected to other domains of the same protein by short sequences without secondary structure.

motility assay: A method for detection of a molecular motor which depends upon observation of nanoscale motion; e.g. the detection of the rotary motion of a fluorophore attached via an actin arm to the -subunit rotating inside the - and -subunit hexamer of an immobilized ATP synthase.

motor: In cell biology, a complex of structures responsible for intracellular movement, e.g. the microtubule-anchored spindle motor, the actomyosin engine. (*see* molecular motor)

MPTP: *see* mitochondrial permeability transition (MPT)

mRNA: Messenger RNA; the RNA that contains the coded information, as sequences of codons, for protein synthesis.

MS/MS: (= tandem mass spectrometry (tandem MS; MS/MS))

msDNA: (= multiple-copy single-strand DNA

MSIA: (= mass spectrographic immunoassay (MSIA))

MS^n: (*see* tandem mass spectrometry (tandem MS; MS/MS))

MSP: (= methylation-specific PCR)

mtDNA: Mitochondrial DNA; a double-stranded polynucleotide composed of a heavy and a light chain, distinguished by the buoyant densities of the separated strands, which are determined by their G+T content.

MTOC: (= microtubule-organizing centre (MTOC))

mucin: An extracellular glycoprotein with a high carbohydrate content that contains many serine and threonine residues bearing the *O*-linked carbohydrate moieties.

mucopolysaccharide: (= proteoglycan (mucopolysaccharide))

mucosa: The external secretory and absorptive surface of the gastrointestinal tract.

multicatalytic proteinase: (= proteasome)

multi-dimensional MS: (*see* tandem mass spectrometry)

multi-drug resistance (MDR): A phenomenon in which resistance that develops to one drug is manifested as resistance to many others, even those with very different structures; seen especially in chemotherapy. MDR is due to traffic ATPase transporters of very low specificity that extrude drugs or sequester them in vesicles.

multienzyme complex: An aggregate of enzymes of specific composition and geometry that act in sequence on a substrate and that, although separable *in vitro*, fit and function together for efficiency, e.g. pyruvate dehydrogenase complex, bacterial fatty acid synthase.

multigene family: A group of genes that code for homologous products with similar functions, e.g. the genes for globins. (*see also* superfamily)

multimeric: Descriptive of a protein composed of several subunits.

multiple-copy single-strand DNA (msDNA): A satellite DNA detected in some bacteria that features an oligoribonucleotide attached by the 2'-hydroxy group of an internal guanylate residue to the 5'-end of an oligodeoxyribonucleotide, while the 3'-end of the DNA moiety is base-paired with the 5'-end of the RNA moiety; both the DNA and RNA possess considerable internal base-pairing.

multiplex amplification: (= multiplex PCR)

multiplex fluoresence in situ hybridization (multiplex FISH): A method involving the use of several fluorophore-tagged oligonucleotides to simultaneously label metaphase chromosomes, and their discrimination and visualization through the use of narrow band filters and computerized output.

multiplex PCR: A screening technique for any of several mutations known to cluster in a relatively small number of sites (hot spots) in a gene, especially microsatellites. Simultaneous PCR reactions with several restriction fragments and the same number of pairs of primers designed to flank the deletion-prone hot spots give as many amplified fragments when normal cells or tissues are scanned, but a mutant will lack one of the restriction fragments.

As the fragments sort into clusters of closely similar sizes, there is no overlap of fragments due to restriction at different sites.

multivesicular body (MVB): A subcellular structure that travels between the inside surface of the plasma membrane and endosomes, perhaps along a tubular endocytotic network. It scavenges endocytosed receptors destined for degradation, delivers them to endosomes and facilitates the return to the plasma membrane of receptors destined for recycling.

multi-wavelength anomalous diffraction (MAD): For proteins that contain a heavy metal, or synthesized with selenomethionine in place of methionine, an approach to solving their X-ray crystallography structure that takes advantage of the anomalous diffraction, in addition to the classical diffraction, that occurs when the crystal is irradiated by synchrotron X-rays of defined energies on the absorption edge of the heavy atoms. The additional diffractions provide sufficient information for assignment of the phases and amplitudes of the crystal's diffraction. (*see also* isomorphous replacement; phase problem)

multiwire data collector: A variety of area detector; an array of individual proportional photon detectors individually wired to a computer to allow quantification of energy and, in crystallographic experiments, its angle of diffraction.

mutagen: A compound that causes a mutation by chemical modification of a base of DNA or as an intercalating agent

mutarotation: The change in optical rotation that accompanies the approach to equilibrium of the two anomers of a sugar from a solution of one anomer alone; also the conversion of one anomer of a sugar into the other.

mutator phenotype hypothesis: A proposal for the mechanism of cancer, based the fact that mutations are relatively much more common in cancer cells than in normal cells, and even increase as a tumour grows. An initial mutation, which creates the "mutator phenotype", allows the accumulation of further mutations and the evasion of the normal checks on the cells growth.

mutein: A mutant protein, especially a product of recombinant DNA technology.

MVB: (= multivesicular body (MVB))

MVR-PCR: (= minisatellite variant repeat-specific priming PCR (MVR-PCR))

MWC model: (= Monod-Wyman-Changeux (MWC) model)

myco-: A prefix derived from the Greek for fungus.

mycoplasma: A type of bacteria that has no cell wall.

myeloma cell: A bone-marrow tumour cell that produces and secretes an immunoglobulin.

myo-: A prefix that indicates muscle, e.g. myocyte.

myoblast: An immature muscle cell.

myokinase: A former name for adenylate kinase.

myoneural junction: (= neuromuscular junction (myoneural junction)).

N

N protein: (*see* anti-terminator)

naive: In immunology, descriptive of an unimmunized animal or a memory T-cell that has not been activated to respond to an antigen.

nano- (n): A prefix which denotes 10^{-9}; e.g. nanomolar, 10^{-9} M.

naptonuon: (*see* junk DNA)

NASBA: (= nucleic acid sequence-based amplification (NASBA))

nascent: Descriptive of a newly formed molecule during its formation or at the moment of its completion.

native gel: A polyacrylamide gel for support of electrophoresis of proteins that is formulated without sodium dodecyl sulphate, so that the protein remains in its native form. (*see also* sodium dodecyl sulphate/polyacrylamide-gel electrophoresis (SDS/PAGE))

native structure: The structure of a protein or nucleic acid in which it is able to perform its physiological function. (*see also* denaturation)

natural killer (NK) cell: A cytotoxic lymphocyte that recognizes foreign cells but, unlike the killer T-cell, need not be previously sensitized to act.

natural product: (*see* secondary metabolism)

natural selection: A hypothetical explanation of evolution, that successful organisms, over time, will crowd out and displace less successful ones, or will be better adapted to different niches and, over time and by genetic drift, lead to new species.

near infrared (NIR) spectroscopy: In biology and medicine, a non-invasive technique for monitoring physiological parameters such as blood flow, cell volume and oxygenation level of redox molecules. NIR radiation penetrates tissues up to several centimetres; transmitted or reflected light of the same wavelength, or scattered light of a longer wavelength (e.g. from a laser Doppler flowmetry device) is collected and analysed.

nearest-neighbour analysis: A method of analysis in which a polymer is first synthesized from a precursor labelled in a specific atom, then degraded in such a way that the labelled atom appears with the adjacent moiety and allows deduction of the identity of the pair that had shared that atom in the polymer.

necrosis: (*see* apoptosis)

negative co-operativity: A form of allosteric behaviour of multimeric enzymes, in which binding of substrate to one subunit decreases the affinity of the substrate for other subunits. (*see also* positive co-operativity)

negative energy balance: The dietary situation in which caloric dietary intake is smaller than energy expenditure, resulting in loss of weight. (*see also* positive energy balance)

N-end rule: The observation that the half-life of a protein *in vivo* is a function of its N-terminal residue. (*see also* trans-recognition)

neoglyco-: A prefix denoting a synthetic complex carbohydrate moiety that has been attached to a protein or lipid.

neoplasm: (= tumour)

nephelometric: Turbidimetric.

Nernst equation: A quantitative expression of the relationship of concentration differences of an ion across a membrane: $=(RT/nF)(\ln[X^n]_e/[X^n]_i)$, where is the membrane potential, R the gas constant, T the temperature, n the ionic charge, F the Faraday constant, $[X]_e$ the external concentration of the ion and $[X]_i$ the internal concentration. Also an expression of the redox potential of a system: $E=E_0+(RT/nF)\ln([ox]^n/[red]^n)$, where E and E_0 are respectively the actual and standard electrical potentials of the system, and [ox] and [red] are the concentrations of the oxidized and reduced species respectively.

nerve gas: Often a fluorophosphate inhibitor of acetylcholinesterase; chemically and functionally related to organophosphate insecticides.

nested deletion sequencing: A method for extension of the length of a target DNA that can be sequenced using a single primer. Removal by an appropriate nuclease of varying polydeoxynucleotide lengths from the region near the primer binding site on a plasmid brings into sequencing range different regions of the target, followed by religation and circularization of the resultant shortened plasmids.

nested primers: A second set of PCR primers whose use follows an initial copying of a nucleic acid sequence; the nested primers are specific for sequences internal to the 3'- and 5'-ends of the originally copied sequence.

Neuberg ester: Obsolete name for fructose 6-phosphate.

neuro-: A prefix derived from the Greek for nerve.

neurofilament: An intermediate filament seen microscopically in neurons that may consist of one of several types of protein.

neuromuscular junction (myoneural junction): The attachment site of a neuron on a muscle cell, analogous to a synapse.

neuron: A nerve cell.

neurophysin: A protein associated with oxytocin or vasopressin in the secretory cells of the neurohypophysis and in blood; derived from the same protein precursor, proneurophysin, as the hormone it binds.

Neurospora crassa: A bread mould; a subject of much experimentation. (*see also* one gene one enzyme hypothesis)

neurotransmitter: A chemical signal that passes from an afferent nerve across the synapse to stimulate the efferent nerve, e.g. acetylcholine, noradrenaline (norepinephrine), dopamine.

neutraceutical: A dietary supplement

neutral glyceride: An uncharged fatty acid ester of glycerol; i.e. a mono-, di- or tri-acylglycerol, as contrasted with a phospholipid.

neutron diffraction: A crystallography technique that uses neutrons (rather than the X-rays of more conventional crystallography) which are scattered by nuclei and can therefore locate hydrogen atoms.

neutrophil: (= polymorphonuclear leucocyte (PMN))

new view: (*see* funnel concept)

nicking: As contrasted with cutting, the cleaving of a single, or very few, phosphodiester bonds in one polynucleotide strand of a DNA duplex; also the cleavage of a peptide bond in a protein that does not denature the protein or release a peptide. (*see also* cut)

nick-translation: A technique often used for introduction of ^{32}P into a DNA duplex involving cleavage of a phosphodiester bond of one strand, excision of the newly exposed nucleoside 5'-phosphate residue, and introduction of a new nucleotide from a radiolabelled triphosphate substrate.

NIH shift: The mechanism of a step in the catabolism of tyrosine, named for the National Institutes of Health, where it was described. The sequence of events by which *p*-hydroxyphenylpyruvate hydroxylase forms homogentisic acid involves a-oxidative decarboxylation to form a peroxyacid, epoxidation of the aromatic ring by the peroxyacid, and migration of the side chain to the site on the ring adjacent to its original position.

ninhydrin reaction: A colorimetric method for quantification of amino acids; reaction with ninhydrin oxidizes and decarboxylates amino acids and generates an intense blue chromophore from reduction and coupling of the reagent with itself.

Nissl substance: (= rough endoplasmic reticulum (RER))

nitrogen balance: The quantitative difference between the intake of all nitrogenous products in the diet and the excretion of all nitrogenous products (both expressed in terms of g of nitrogen); *positive nitrogen balance* if there is a net uptake of nitrogen and *negative nitrogen balance* if a net loss.

nitrogen cavitation: A technique for disruption of cells prior to their fractionation. A suspension of cells is subjected to high nitrogen pressure, which is suddenly released so as to cause the cells to burst as nitrogen bubbles form inside them.

nitrogen fixation: The reduction of molecular nitrogen to ammonia and nitrogen-containing metabolites that is carried out synthetically, or naturally by blue-green algae or bacteria, e.g. *Rhizobium* species that inhabit nodules on the roots of legumes.

nitrogen mustard: An alkylating agent that can cross-link adjacent guanine bases of DNA and thus interfere with its function, e.g. $CH_3N(CH_2CH2Cl)_2$ (mechlorethamine).

nitrous acid method: (*see* Van Slyke method)

NK cell: (= natural killer (NK) cell)

N-linked carbohydrate: A carbohydrate moiety of a glycoprotein that is attached by a glycosidic bond to the -amide nitrogen of an asparagine residue; may be complex-type or high-mannose-type. (*see also* O-linked carbohydrate)

NMR: (= nuclear magnetic resonance (NMR))

noci-: A prefix from the Latin for hurt, which denotes a harm.

node: The point of contact of a strand of supercoiled DNA where it loops back on itself. A node, and the *writhe* of the DNA, may be positive or negative; it is positive if, when viewed so that the strand enters the node from below and leaves from above, and the shortest turn from strand entry and strand exit is clockwise. Negative nodes are formed from negatively supercoiled DNA. A *knot* in the DNA is seen when three nodes all of the same sense, e.g. all negative, occur together; this is seen in electron micrographs as a *trefoil* structure, a knot from which extend three DNA loops. (*see also* intasome)

NOE: Nuclear Overhauser effect. (*see* nuclear Overhauser and exchange spectroscopy (NOESY))

NOESY: (= nuclear Overhauser and exchange spectroscopy (NOESY))

Nomarski interference optics: A microscope optical system for visualization of differences in refractive indices so as to observe unstained specimens.

nomogram: The presentation of three parallel linear or non-linear scales of related values aligned so that a straight line between desired points on two of the scales also crosses the third scale at the corresponding value, e.g. a line through points on scales for centrifuge rotor diameter (cm) and speed (rev./min) also passes through the scale for centrifugal force (g) at the correct point.

non-coding strand: (*see* antisense)

non-collagen collagen: A protein complex that, like collagen, has (Xaa-Yaa-Gly)$_n$ amino acid sequences and non-helical globular regions. The collagen-like sequences assemble together into a triple helix and the globular regions project out in a '*flower bouquet*' pattern; e.g. acetylcholinesterase, complement protein C1g.

non-competitive inhibition (mixed inhibition): A form of enzyme inhibition in which the inhibitor binds to both the free enzyme and the enzyme-substrate complex, resulting in an increase in K_m and a decrease in V_{max}. (*see also* competitive inhibition; inhibitor; uncompetitive inhibition)

non-essential amino acid: (*see* essential amino acid)

non-haem iron protein: (= iron-sulphur protein)

non-overlapping: Descriptive of the genetic code, in which the three bases of one codon are distinct from the bases of adjacent codons.

non-permissive host: A cell that does not allow the multiplication within it of a virus. (*see also* permissive host)

non-productive: In enzymology, descriptive of an enzyme-substrate complex other than the Michaelis complex; one that cannot be converted into an enzyme- product complex. In protein chemistry, descriptive of a partially folded conformer that is not on a folding pathway between a random coil and the native state. (*see also* framework model)

non-reducing sugar: A sugar whose anomeric carbon atom forms part of a glycosidic bond, rendering it unable to be oxidized by an alkaline cupric ion solution. (*see also* reducing sugar)

nonsense mutation: A mutation that generates one of the three chain termination codons. (*see also* amber mutation; ocher mutation; opal mutation)

non-synonymous: (*see* synonymous)

non-transcribed strand: (*see* antisense)

nor-: A prefix that signifies a product of the removal of a -CH_2- or CH_3 group, e.g. noradrenaline, a 19-norsteroid.

normalized cDNA library: A library that includes all cDNA clones at approximately equal frequencies, unlike most cDNA libraries, which represent cDNAs in rough proportion to the occurrence of their mRNA in the source.

northern blotting: By analogy with Southern blotting, a technique for detection of specific RNA molecules; RNA from a cell is denatured and separated by slab gel electrophoresis, then blotted on to a sheet of nitrocellulose or nylon and hybridized with radiolabelled DNA that is complementary to the desired RNA, whose presence is subsequently indicated by autoradiography. (*see also* electroblotting; Southern blotting; south-western blotting; western blotting (immunoblotting))

nr: A prefix signifying a non-radioactive procedure, e.g. nrPTT

NSF: *N*-ethylmaleimide-sensitive soluble fusion protein. (*see* SNARE hypothesis)

N-terminal: In a polypeptide sequence, that unique residue which is connected to the linear sequence by its carboxy group, leaving it with a free amino group. In practice, the amino group of an N-terminal residue may be modified, e.g. by acylation by an acetyl or fatty acyl group. (*see also* C-terminal)

N-terminus: (*see* C-terminus)

nt: Nucleotide; an abbreviation of the measure of chain length of a polynucleotide

NTP: Nucleoside triphosphate; the letter N refers to any or all of the common bases.

nuclear envelope: The structure that encloses the eukaryotic nucleus and consists of an inner and an outer nuclear membrane and the perinuclear space between the two.

nuclear localization sequence (NLS): *see* nuclear pore complex

nuclear localization signal: A polypeptide sequence of a protein that permits its entry through nuclear pores into the nucleus of the cell.

nuclear magnetic resonance (NMR): A technique for studying the microenvironment of certain atoms that can absorb energy in a magnetic field, e.g. ^{1}H, ^{13}C, ^{15}N, ^{31}P. NMR can be used to study the conformation in solution of polypeptides, proteins and other molecules. (*see also* transverse relaxation-optimized spectroscopy)

nuclear overhauser and exchange spectroscopy (NOESY): A nuclear magnetic resonance method for evaluation of internuclear distances [within 4Å (0.4nm)] between non-covalently bonded nuclei; one centre is saturated with radiation and, as its excited state decays, the effect on the spin distribution of the neighbouring centre (the *nuclear Overhauser effect*) is observed. (*see also* correlation spectroscopy)

nuclear pore complex (NPC): The site on the nuclear membrane through which proteins are transported into, and RNA is transported out of the cell nucleus. The NPC includes proteins that recognize nuclear localization sequences (NLSs), amino acid sequences that mark proteins destined for import into the nucleus, and the GTP hydrolysis-dependent mechanism that drives transport. The complex of the transported protein and the components of the NPC which bind it are termed the nuclear transport complex (NTC).

nuclear pore: A gap in the nuclear envelope that provides a selective barrier to entry or exit of macromolecules.

nuclear receptor: A nuclear protein that allows regulation of gene expression by a lipophilic effector molecule, such as a steroid or thyroid hormone, eicosanoid, retinoid or ecdysone. When the effector molecule is bound to the receptor protein, the latter can also bind to a specific DNA sequence. (*see also* cytosolic receptor)

nuclear run-off assay: A method to determine which genes in a population of cells are expressed at a given time. Nascent RNA transcripts are radiolabelled with [^{32}P]NTPs as they are elongated in isolated nuclei (new transcripts are not initiated in these isolated nuclei). To determine if a specific gene is expressed, the radiolabelled run-off transcripts are used as hybridization probes with DNA from recombinant plasmids that contain the gene of interest. (*see also* run-on assay)

nuclear scaffold: The network of protein fibres that underlies the nuclear inner membrane and is exposed upon extraction of soluble proteins from the nucleus. Between mitoses (interphase), chromatin is attached to the scaffold at points along its DNA strands. The loops of unattached DNA are presumably available for transcription.

nuclear transport complex (NTC): (*see* nuclear pore complex)

nucleation: In the process of folding of a protein, the rate-limiting formation of the first elements of secondary or tertiary structure around which the remainder of the protein subsequently folds.

nucleic acid sequence-based amplification (NASBA): A method for amplification (up to 10^9-fold) of a single-stranded RNA template. A reaction mixture contains RNA polymerase, ribonuclease H, reverse transcriptase, nucleoside triphosphates and two polydeoxynucleotide primers; primer 1 contains a promoter for the RNA polymerase and a sequence complementary to the 5'-end of the RNA template's coding region, and primer 2 is the DNA analogue of the 3'-end of the template. The amplification begins as primer 1 anneals to the template, the reverse transcriptase forms a cDNA:RNA hybrid, the RNA of which is hydrolysed by ribonuclease H. With formation of the resulting single-stranded (ss)DNA, the reaction enters a cyclic phase: the ssDNA anneals to primer 2, and the reverse transcriptase then synthesizes a complementary copy of the first-synthesized DNA to generate a double-stranded promoter region which the polymerase recognizes and uses to make 10-100 RNA transcripts. As each new RNA transcript is a template for the reverse transcriptase, the cycle continues to amplify the original RNA template.

nucleic acid: DNA or RNA; a macromolecule formed of repeating nucleotide units linked by phosphodiester bonds.

nucleo-: A prefix derived from the Latin for kernel, which in physics and chemistry refers to the core of the atom, and in biochemistry and biology to the large structure in interior of most cells.

nucleofuge: A chemical group that is displaced in a nucleophilic substitution or elimination reaction. (*see also* electrofuge)

nucleolus: A structure in the nucleus of a cell that is the site of synthesis of rRNA.

nucleomorph: A cellular inclusion, bound by multiple membranes, that contains a small amount (ca. 500 kbp) of genomic DNA. These inclusions are thought to be the products of secondary endosymbiosis, e.g. in an ancient primary endosymbiotic event, a photosynthetic prokaryote was engulfed by a heteroptrophic alga; the resulting photosynthetic alga later was itself engulfed by another heterotroph. The nucleomorph retains its photosynthetic capacity, and as a memento of its provenance, the membranes which surround it.

nucleophile: A compound or functional group that can attract and bind a proton or another positively charged species such as a carbocation ion. (*see also* electrophile)

nucleoplasm: Between mitoses (interphase) the visibly unstructured matrix of the nucleus, which consists of all but the nucleolus.

nucleoprotein: A protein-DNA or protein-RNA complex.

nucleoside: A purine or pyrimidine base with a sugar attached in a glycosidic linkage; *ribonucleoside* if the sugar is D-ribose and *deoxyribonucleoside* if the sugar is D-deoxyribose. (*see also* nucleotide)

nucleosome: The repeating structural unit of chromatin that consists of a complex of eight molecules of histones and a DNA double helix wrapped twice around them.

nucleotide: A nucleoside with one, two or three phosphate(s) esterified to one of the free hydroxy groups of the sugar moiety, usually the 3'- or 5'-hydroxy group; *ribonucleotide* if the sugar is D-ribose; *deoxyribonucleotide* if the sugar is D-deoxyribose. (*see also* nucleoside)

nucleotide-binding domain: One of the homologous supersecondary structures of some proteins, composed of Rossmann folds, to which may be bound the AMP moiety of a substrate or ligand; e.g. the domain of a dehydrogenase that binds a pyridine nucleotide or a flavin-adenine dinucleotide.

nucleotide-binding fold: (= nucleotide-binding domain)

nucleus: A membrane-limited structure of eukaryotic cells that contains chromatin and is the site of DNA and RNA synthesis.

nudix hydrolase: *see* housecleaning enzyme

nuon: (*see* junk DNA)

nut site: N utilization site. (*see* anti-terminator)

O

-oma: Suffix from the Latin indicating a structure or mass.

-ome: = -oma

O: 'Farrell gel (= two-dimensional electrophoresis)

O-antigen: (*see* lipopolysaccharide (LPS))

O-linked carbohydrate: The carbohydrate chains that are attached by a glycosidic bond to a protein through the hydroxy group of a serine or threonine residue. (*see also* N-linked carbohydrate)

occupancy: The degree to which a binding site, e.g. a receptor, is liganded, e.g. by its hormone.

ocher mutation: A nonsense mutation that, due to the premature appearance in the mRNA of the terminator codon UAA, leads to the formation of a truncated, possibly non-functional, protein. (*see also* amber mutation; opal mutation)

ODN: Oligodeoxynucleotide

oestrogen (estrogen): A compound, usually a steroid, that supports the development of female secondary sex characteristics; e.g. oestradiol.

oestrogenic chemical: = endocrine disrupter

off-pathway intermediate: *see* paradox of Levinthal

ogston hypothesis: The proposal that an enzyme can distinguish in its substrate between two like substituents on a carbon atom that also carries two unlike groups, through binding sites for three of the four substituents, which forces the substrate to display in a unique configuration one of the like groups. (*see also* chirality; meso-carbon; symmetry)

Okazaki fragment: Fragments of a single-stranded DNA synthesized on the discontinuous site of a DNA replication fork. (*see also* semi-discontinuous)

OLA: (= oligonucleotide ligation assay)

old view: *see* funnel concept

oligo-: A prefix that indicates a moderate degree of polymerization, as in oligonucleotide, oligopeptide, oligosaccharide. (*see also* mono-; oligomer; poly-)

oligomer: A small polymer of complexity greater than that of a monomer but less, in common usage, than that of a dodecamer.

oligomerization: (*see* aggregation)

oligonucleotide array: (= chip (oligonucleotide array))

oligonucleotide ligation assay (OLA): A method to identify a specific oligonucleotide sequence without use of radiochemicals, electrophoresis or centrifugation. An oligonucleotide biotinylated at its 5'-end and another with a reporter group (chromophore or fluorophore) at its 3'-end are constructed to hybridize to the sequence to be detected in a template DNA strand. A ligase is able to join the two oligonucleotides if they both match the template perfectly. The ligated oligonucleotide can be extracted from the mixture on to immobilized streptavidin and the presence of the reporter group detected.

oligosaccharin: One of a group of complex oligosaccharides shown to have a regulatory function in plants, i.e. growth regulation, organogenesis, defence against pathogens.

omega-loop: An irregular secondary structure on the surface of many globular proteins that consists of 6-16 amino acid residues folded into a rigid, tightly packed loop in which the N- and C-terminal ends are located close together, thus forcing the sequence into an -shaped feature.

OMH: (= optimal matching hydrophobicity)

omic research: Following the currency of 'genomics', additional words have been coined to denote the compilation of data bases as resources for future research and analysis; proteonomics for proteins, kinomics for protein kinases, CHOnomics for carbohydrates, transcriptomics, vaccinomics, clonomics, functional, structural and pharmacogenomics, etc. Together, these approaches are termed omic research.

oncogene: A gene present in normal cells which when altered can transform the normal cell into a malignant cell. (When emphasizing this relationship, the unaltered gene is called a proto-oncogene and the altered gene, an oncogene.) The alteration may result in overproduction of a gene product or faulty function. An oncogene is denoted by the prefix c if a cellular oncogene, e.g. c-*ras*, or by the prefix v if a viral oncogene, e.g. v-*ras*, which is presumed to have originated from a capture of the c-oncogene by the virus during an infection of a normal cell. Contrasted with a *thanatogene*, or *death gene*, which when activated leads to apoptosis. (*see also* anti-oncogene (tumour suppressor gene); oncoprotein)

oncoprotein: The translation product of an oncogene; a regulator of cellular processes, which if the gene is mutated, results in uncontrolled growth (cancer). Oncoproteins may be embedded in the plasma membrane and serve as receptors, reside in the cytosol and be involved in signal transduction, often associated with a tyrosine kinase activity, or may be found in the nucleus and be directly involved in gene regulation. (*see also* oncogene)

one-by-one type mechanism: (*see* zipper-type mechanism)

one-carbon pool: The aggregate of freely interconvertible moieties that contain single carbons more reduced than carbon dioxide; i.e. metabolites of tetrahydrofolate, the methyl groups of choline, the hydroxymethyl of serine, etc.

one-electron carrier: A compound that can accept or donate a single electron; e.g. a cytochrome.

one-gene one-enzyme hypothesis: The proposal that arose from genetic studies on the effects on amino acid metabolism of mutations in *Neurospora*, that there is a one-to-one correspondence between enzymes and the genes that encode them.

one-hybrid system: (*see* two-hybrid system)

on-pathway intermediate: (*see* paradox of Levinthal)

oo-: A prefix derived from the Greek for egg.

oocyte: A female gamete, a precursor of an ovum.

opal mutation: A nonsense mutation that, due to the premature appearance in the mRNA of the terminator codon UGA, leads to the formation of a truncated, possibly non-functional, protein. (*see also* amber mutation; ocher mutation)

open reading frame (ORF): One of three possible reading frames in which an mRNA is potentially translated into protein. In analysis of a DNA sequence, an ORF is characterized by the sequence of nucleotides that when transcribed into mRNA results in a series of triplet codons that is not interrupted by a translation termination codon. (*see also* unassigned reading frame (URF))

open-chain form: The form of a sugar in which there has been no internal condensation to form a pyranose or furanose. (*see also* anomer; alpha ()-isomer; beta ()-isomer)

operator: A locus on DNA that controls transcription when a repressor or activator becomes bound.

operon: A group of genes, usually metabolically related, that can be controlled co-ordinately from a common regulatory sequence. (*see also* cistron)

opiate: Morphine, pharmacologically active derivatives of it or compounds such as enkephalins and endorphins that similarly act at the same receptors in the brain.

opsonin: An agent of the blood, especially of the complement system, which can react with by-products of cell damage and with foreign substances and particles, such as bacteria. Opsonization allows leukocytes and macrophages to recognize and endocytose these substances and particles. *see* innate immunity.

optical rotation: The change in the angle of transmitted plane-polarized light as it passes through a solution of a chiral sample, such that the change is proportional to the concentration of the chiral molecule and to the length of the light path through the solution. In *dextro*-rotation the change is clockwise when facing the light source; in *laevo*-rotation the change is counterclockwise. (*see also* circular dichroism (CD); optical rotatory dispersion (ORD))

optical rotatory dispersion (ORD): The optical rotation of a material as a function of wavelength. (*see also* circular dichroism (CD))

optimal matching hydrophobicity (OMH): A calculated value used to predict conformational similarities of proteins from their amino acid sequences. Numerical values for hydrophobicity, OMHs, are assigned to each amino acid residue to allow comparisons based upon features of the residues that are important to their structural function, but independent of actual amino acid identities. An OMH plot, OMH values against sequence number, permits visual comparisons of proteins. (*see also* hydropathy index; hydropathy plot)

optode: A biosensor with a light-detecting sensor, e.g. a device that detects the lowering of pH due to the action of glucose oxidase at the end of a fibre optical light guide by the effect on a pH-sensitive fluorescent dye.

orbital perturbation theory: (*see* entropy effect)

orbital steering: (*see* entropy effect)

ORC: (= origin recognition complex)

ORD: (= optical rotatory dispersion)

ordered: (*see* enzyme mechanism)

ORF: (= open reading frame)

organ targeting: The homing of cells or viruses to specific tissues or organs, which is facilitated by organ-selective address molecules on the target tissue surface that tightly bind to cell-specific markers on the cell or virus surface.

organelle: Inclusion in the cell cytoplasm that can be membrane-limited (e.g. chloroplast, mitochondrion, endoplasmic reticulum, Golgi apparatus) or non-membrane-limited (e.g. cytoskeletal element, ribosome, nucleosome).

origin of replication: A double-stranded DNA sequence that binds the proteins that will begin unwinding the helix, preparatory to initiation of replication.

origin recognition complex (ORC): An aggregate of nuclear proteins involved in initiation of replication and/or repression of transcription (silencing). The DNA regions that bind the ORC are examples of *anomalously replicating sequences* (*ARSs*), which promote plasmid replication in yeast. Other examples are *silencers*, which flank repressed genes; the sequences that interact directly with an ORC are *ARS consensus sequences* (*ACSs*).

orphan receptor: A membrane protein that has been discovered by cloning of its gene to be of the same family as known receptors, e.g. the G-protein-coupled receptors, but for which the activating ligand has not been identified. The process of identification of the natural ligand for orphan receptors has been called 'reverse physiology'.

orphon: A displaced genetic element; a coding or non-coding sequence that usually occurs in a series of tandem repeats in a eukaryotic genome but appears in a location atypical of the species or strain.

ortho-: A prefix derived from the Greek for straight. In aromatic compounds, it signifies the relation of substituents on adjacent nuclear carbon atoms. (*see* meta-)

orthologous: Descriptive of homologous genes that have evolved divergently after speciation, e.g. cytochromes *c* of various species. (*see also* paralogous)

orthologue: (*see* cluster of orthologous groups)

osazone: The product of coupling the carbonyl carbon of a sugar with phenylhydrazine, its further oxidation and subsequent addition of a second phenylhydrazine to form a crystalline compound.

osmolarity: The concentration of non-permeable solutes that contribute to osmotic pressure; *iso-osmotic* or *isotonic* if equal to that of a cell, *hypotonic* if lower and *hypertonic* if higher.

osmotic pressure: The pressure developed by a solution of a non-permeable solute isolated by a membrane from a surrounding solution.

osmotolerant: Descriptive of organisms that may survive extremes of salt or non-ionic solute concentrations. (*see* halophilic)

osteo-: A prefix that indicates bone, e.g. osteocyte.

Ouchterlony plate: (*see* immunodiffusion)

outer mitochondrial membrane: (*see* mitochondrion)

outer nuclear membrane: (*see* nuclear envelope)

overhang: The extension of one polynucleotide strand beyond the terminus of its complementary strand.

Overhauser effect: (= nuclear Overhauser effect; *see* nuclear Overhauser and exchange spectroscopy (NOESY))

overlapping gene: A rare phenomenon found in small viral genomes whereby a single mRNA can be translated in two different reading frames to produce two different proteins.

ovine: Derived from sheep

ovum: A female gamete, which can be fertilized by a spermatozoon.

oxidation-reduction potential: (= redox potential ($E^{O'}$))

oxidative decarboxylation: Removal of carbon dioxide from a carboxylic acid that is facilitated by oxidation of the -carbon (*-oxidative decarboxylation*, e.g. the pyruvate dehydrogenase reaction) or the -carbon (*-oxidative decarboxylation*, e.g. the isocitrate dehydrogenase reaction).

oxidative phosphorylation: The reactions that synthesize the phosphoanhydride bond of ATP by the coupling of that energetically unfavourable reaction with the spontaneous oxidation of metabolites. (*see also* chemiosmotic theory; P:O ratio)

oxidative stress: The cumulative damage due to the less than 100% effectiveness of antioxidants in prevention of free radical reactions such as lipid peroxidation.

oxidoreductase: One of a class of enzymes that oxidize one substrate as they reduce another, e.g. a dehydrogenase, an oxidase, a peroxidase.

oxygen dissociation curve: A characterization of an oxygen-binding protein, such as haemoglobin; a plot of percentage saturation against partial pressure of oxygen.

P

-phage: A suffix derived from the Greek for eating; e.g. macrophage.

-plasia: A suffix derived from the Greek for formation.

-plast: A suffix from the Greek for formed, indicating an organized particle or cellular inclusion; e.g. chloroplast.

-poiesis: A suffix derived from the Greek for creation and which refers to a process of creation or development, e.g. megakaryocytopoiesis.

p: In addition to use as the metric notation of pico (10^{-9}), it is commonly used to indicate the logarithm of a reciprocal, e.g. $pK_a=\log(1/K_a)$. When seen as a prefix followed by a number, p indicates molecular mass determined from a Laemmli gel and characterizes a protein whose identity is otherwise unknown, e.g. p53, a protein of mass 53000Da. It is also used as a prefix to an alphanumeric designation of a plasmid, e.g. pVA50. In nucleic acid chemistry it denotes an orthophosphate in a 5'- or 3'-phosphomonoester bond (e.g. pA or Gp), a phosphodiester bond (e.g. CpG), or a phosphate anhydride bond (e.g. ppA). In genetics it is a designation of a location in the short arm of a chromosome. (*see also* centromere)

P:O ratio: The ratio of phosphate incorporated into ATP to oxygen atoms reduced to water; a measure of the efficiency of coupling of phosphorylation to oxidation. Based on the now-obsolete chemical coupling theory of oxidative phosphorylation, the passage of electrons down the electron transport pathway results at several points in the chemical synthesis of ATP. The number of phosphates that are fixed into ATP for every two electrons that pass from a substrate to reduce each oxygen atom of molecular O_2 (the P:O ratio) corresponds to the number of such points and should be an integer value. Experimentally, it was found to be between 2 and 3 when NADH was the donor of electrons and between 1 and 2 when succinate was the donor, and the theoretical integer values were set to 3 and 2 respectively. The chemiosmotic theory similarly predicts *H^+:O* and *H^+:ATP ratios*. Experimentally these appear to be 10 and 4 respectively when NADH is the substrate, equivalent to a P:O ratio of 2.5, and 6 and 4 respectively for FAD-linked substrates (e.g. succinate), equivalent to a P:O ratio of 1.5.

packaging cell: In biotechnology, a eukaryotic cell that is used to produce a modified virus.

PAD: (= pulsed amperimetric detection (PAD))

PAGE: Polyacrylamide-gel electrophoresis. (*see* electrophoresis; native gell; sodium dodecyl sulphate/polyacrylamide-gel electrophoresis (SDS/PAGE))

palindrome: A concept borrowed from linguistics; a segment of a double-stranded polynucleotide such that the order of bases, read 5'-to-3' in one strand, is the same as that in the complementary antiparallel strand, read 5'-to-3'; a common structural feature of DNA sequences that constitute specific protein-binding sites.

pan editing: (*see* guide RNA (gRNA))

pan-: A prefix derived from the Greek for all.

para-: A prefix derived from the Greek for past or beyond. In benzene compounds, para- signifies the relation of substituents on nuclear carbons which are opposite each other. Elsewhere, para- signifies modification.

paracellular: (*see* tight junction)

paracrine: Descriptive of a secretion that acts on cells adjacent to the site of its secretion. The term is applied to the non-synaptic neurotransmitter-mediated communication between cells of the central nervous system. (*see also* autocrine; endocrine; exocrine; telecrine function)

paradigm: In science, a key example that represents and illustrates a range of related cases; e.g. Lac as a paradigm of bacterial operons.

The word is also used as a synonym for model, such as a knock-out or transgenic animal.

paradox of Levinthal: The discrepancy between the enormous number of conformations a polypeptide may assume and the rapidity with which it normally achieves its native conformation. It excludes the possibility that each conformation is sampled and only the thermodynamically most stable form persists and proposes a pathway that leads denatured conformations to the native state. This proposal was later considered naïve, and has been superceded by the funnel concept. (*see* framework model)

parallel (or combinatorial) synthesis: (*see* small molecule library)

paralogous: Descriptive of homologous genes that have duplicated and evolved divergently, e.g. those encoding trypsin, chymotrypsin, elastase and thrombin, which are all serine proteinases and occur together within the same species. (*see also* orthologous)

paralogue: (*see* cluster of orthologous groups)

paramagnetic: Descriptive of a chemical species that can interact with a magnetic field, usually a transition metal ion or an organic compound with an unpaired electron. (*see also* diamagnetic)

paranemic: (*see* plectonemic)

paratope: The antigen-binding site of an antibody. (*see also* epitope)

parenchyma: The functional tissue of an organ, as contrasted with the stroma.

parenteral nutrition: Provision of food other than orally, usually by intravenous infusion.

PARF: (= polymorphic amplifiable restriction endonuclease fragment (PARF))

partition chromatography: A technique for separation of molecules based on their differing ratios of solubility in two immiscible solvent phases. A stationary phase is coated or impregnated with one solvent phase and the mixture to be separated is passed over it in the mobile phase.

partition coefficient: The ratio of solubility of a substance in two immiscible solvents, e.g. in oil and water.

partner: In cell biology, one of a pair of macromolecules which bind to create intracellular or membrane-associated complexes, or extracellularly as cell adhesion factors.

PAS stain: (= periodic acid-Schiff (PAS) stain)

passive diffusion: The transport of compounds across a membrane that is unmediated by any mechanism and is independent of energy sources, and therefore occurs at a rate that is determined by the area of the membrane, the concentration difference across it and the solubility of the compound in the membrane. (*see also* active transport)

patch-clamp technique: A method for measurement of conductance of cell membranes by clamping a tiny buffer-filled pipette tip to a cell surface, so that the pipette serves as an electrode for the measurements.

patched circle PCR: A method for site-directed deletion of a large-scale fragment from a cloning vector, e.g. for obtaining cDNA from a plasmid without vector sequences. Two PCR primers pair to the 5'- and 3'-flanking sequences adjacent to and extending away from the unwanted sequence, and the DNA is amplified by PCR. Another oligonucleotide is added to bind to both ends of the amplified linear cDNA and patches them together to mimic closed circular DNA, which is then used to transform *E. coli.*

patching: The migration of receptors to certain areas on the surface of cells when exposed to ligands, especially the interaction of an antibody with its receptors on the surface of lymphoid cells; also called *capping.*

pathway: (= metabolic pathway)

paucidisperse: (*see* monodisperse)

PCD: Programmed cell death. (= apoptosis)

PCR: (= polymerase chain reaction)

PCR amplification of multiple specific alleles (PAMSA): A method for determination of zygosity. Three primers are used, two of which are allele-specific and one of which matches the complementary strand of both alleles. Allele specificity is achieved by designing mismatches to the unwanted allele at the 3' end of the primer. However, precisely because the primers must be chosen to generate amplicons of characteristic size, they will have different efficiencies of amplification, which will cloud the interpretation of results. (*see* bidirectional PCR amplification of specific alleles; PCR amplification of specific alleles)

PCR amplification of specific alleles (PASA): PASA; a method for genotyping of single-base mutations. Genomic DNA is used as a template in two reactions, one with a primer that makes a perfect match to the wild-type allele, another with a primer matched to the mutated allele; both reactions contain a second primer that makes a perfect match to the opposite strand of either wild-type or mutant allele. PCR will preferentially amplify fragments from an allele which is present in the genomic DNA. Characteristic results will be seen for wild-type and mutant homozygotes and for heterozygotes.

PCR-based differential scanning: A method for amplification of a heterogeneous population of cDNAs preparatory to creation of a library. Total mRNA is copied first using an oligo(dT) primer to obtain one copy, then that primer is removed and the polynucleotides are tailed with oligo(dC) so that the second strand can be synthesized using an oligo(dG) primer. This second strand can then be amplified by PCR using oligo(dT) and oligo(dG) primers.

PD-MS: Plasma-desorption mass spectrometry. (*see* mass spectrometry (MS))

P-DNA: A souble-stranded form of the nucleic acid which differs from B-, A- and Z-DNA in being less tightly coiled and more extended and having the phosphate backbone on the outside.

pellet: The sedimented portion that accumulates during centrifugation. (*see also* supernatant fluid)

pentacovalent intermediate: An intermediate or transition state in phosphate reactions, in which an incoming oxygen atom displaces a leaving oxygen in a concerted reaction and forms a *trigonal bipyramid* in which the phosphorus atom is at the centre, three oxygens are co-planar around it and the entering and leaving oxygens are normal to the plane on opposite sides of the phosphorus atom.

pentose nucleic acid: Obsolete name for RNA.

pentose phosphate pathway: The enzymic reactions that oxidatively convert glucose 6-phosphate into ribulose 5-phosphate and then to intermediates of the glycolytic pathway. The two oxidative reactions are the major source of NADPH in many cells. Also known as the Dickens-Warburg pathway, the hexose monophosphate shunt, the pentose shunt and the phosphogluconate oxidative pathway.

pentose shunt: (= pentose phosphate pathway)

penultimate: Descriptive of a position in a linear sequence, e.g. of a polypeptide, adjacent to one of the terminal positions.

peptidase: An enzyme that cleaves the peptide bonds of proteins and peptides. (*see also* endopeptidase; exopeptidase).

peptide: A compound formed by incomplete hydrolysis of a protein, or by elimination of the elements of water from between the -carboxy and the -amino groups of -amino acids to form a linear polymer.

peptide bond: The amide bond formed by condensation of the -carboxy group of one amino acid with the -amino group of another; the bond that joins together the amino acid residues that comprise a protein, peptide or polypeptide.

peptide site: The part of a ribosome that binds the growing peptidyl-tRNA before the peptidyl group is transferred to the next amino acyl residue which is held at the amino acyl site as its tRNA ester.

peptide Velcro: A description of helix-helix interactions in which side chains of two parallel, adjacent helices closely interact due to complementary shapes and electrostatic interactions, e.g. the leucine zipper motif of the fos and jun heterodimers.

peptidoglycan: A polymer of bacterial cell walls that exhibits considerable species variation in structure. It is generally composed of a heterodisaccharide attached to a peptide; the carbohydrate moieties of neighbouring subunits are cross-linked, and the peptides of neighbouring subunits are also cross-linked.

peptidomimetic: A small non-peptide molecule, often cyclic, with features of a natural, biologically active peptide, which confer upon it the natural peptide's activity, but which is more stable or easier to manufacture.

peptidyl: Descriptive of a peptide as it is attached by its -carboxv group, e.g. a peptidyl-tRNA.

peptoid: An alalog of a polypeptide, a poly-glycine in which the backbone nitrogen atoms are substituted with groups the same as or different from those of natural amino acids. Although these polymers do not have chiral centres, their N-substituted amide bonds are set into cis or trans conformations, which allows them to adopt helical structures, somewhat like -helices.

per-: A prefix derived from the Latin for in accord, which indicates completeness.

PER: Pre-edited region. (*see* guide RNA (gRNA))

perfusion: An *in vitro* or *in situ* technique to circulate a fluid through the blood vessels of an organ. (*see also* perifusion)

perfusion chromatography: A variant of high-pressure liquid chromatography in which particles as the stationary phase have pores large enough [6000-8000Å (600-800nm)] to permit solvent flow-through and thus allow much faster flow rates, decreased diffusional spreading and enhanced access to the full surface of the stationary phase.

peri-: A prefix derived from the Greek for around.

perifusion: An *in vitro* technique to circulate fluid around an isolated organ or tissue held in a specially designed chamber. (*see also* perfusion)

perinuclear space: The space between the inner and outer nuclear membranes.

periodic acid-Schiff (PAS) stain: A reagent for staining of carbohydrate in polyacrylamide gels or histological preparations; periodic acid first oxidizes vicinal hydroxy groups to generate aldehydes that can then condense with fuchsin sulphurous acid (Schiff's reagent) to form an intensely coloured Schiff base.

peripheral nervous system (PNS): A division of the nervous system of higher animals that consists of the neurons that extend from the spinal cord to skeletal muscles, glands, etc. and the neurons that return from these structures to the spinal cord. (*see also* central nervous system (CNS))

peripheral protein (extrinsic protein): A protein loosely associated with a membrane. (*see also* integral protein (intrinsic protein))

peripheral tissue: In any particular discussion, tissues that are not of central concern; e.g. in urea synthesis, tissues other than liver; in digestion, tissues other than the gastrointestinal tract.

periplasmic space: The space between the cell membrane and the outer cell wall of some non-animal and prokaryotic cells.

permeability coefficient: A quantitative measure of the rate at which a molecule can cross a membrane such as a lipid bilayer; expressed in units of cm/s and equal to the diffusion coefficient divided by the width of the membrane.

permease: An active transport mechanism.

permissive host: A cell in which a virus is able to replicate. (*see also* non-permissive host)

permissive temperature: A temperature at which a temperature-sensitive mutant will grow.

peroxidase: A haem enzyme that abstracts two hydrogens from one substrate to reduce hydrogen peroxide, its other substrate, to water.

peroxisome (microbody): A vesicle containing various oxidizing enzymes (which generate H_2O2) and catalase (which degrades H_2O2).

Perrin plot: A graphical method to determine limiting anisotropy or limiting polarization; a display of the reciprocal of polarization against the ratio of absolute temperature to viscosity. The anisotropy or polarization is measured at increasing viscosities (effected by the addition of a solute such as sucrose or glycerol) or decreasing temperatures and extrapolated to absolute zero or infinite viscosity.

PERT: (= phenol-emulsion reassociation technique (PERT))

PEST hypothesis: The proposal that polypeptide sequences rich in the amino acid residues proline (P), glutamate (E), serine (S) or threonine (T) render a cellular protein susceptible to rapid turnover.

petite mutation: Mutation observed in a strain of yeast that is deficient in mitochondrial protein synthesis. The mutants were instrumental in the discovery of mitochondrial DNA and mitochondrial protein synthesis.

pH: A measure of acidity of aqueous solutions; the negative of the common logarithm of the hydrogen ion concentration, $-\log[H^+]$. Values below 7 are acidic; values above 7 are alkaline.

pH stat: A device that automatically delivers and records against time the volume of acid or alkali necessary to maintain a preset pH during a reaction that consumes or generates acid.

phage: (= bacteriophage)

phage artificial chromosome (PAC): *see* bacterial artificial chromosome

phage-display: A method for incorporation, into the DNA that encodes a phage coat protein, of a polypeptide, where it is available for a selection procedure; especially for production and selection of antibody (even anti-self antibody) fragments directed against a specific antigen. A library of V_H or V_L genes is produced by PCR using primers directed to constant regions that flank the variable region, cloned, and inserted into the gene for a phage coat protein that, when expressed, is available for cycles of growth and selection by adsorption on to an antigen-coated plate or beads.

phagemid: A multifunctional cloning vector; a chimaera of a plasmid and a bacteriophage that incorporates and consolidates many useful features of both.

phago-: A prefix with the same derivation as -phage

phagocytosis: (*see* endocytosis)

phagolysosome: A vesicle formed by fusion of a phagosome with a lysosome.

phagosome: A vesicle that forms in a cell as a result of phagocytosis and that contains the engulfed material.

pharmacophore: In the drug design, the minimum of structural features of a candidate drug that will allow binding to the target protein.

pharming: (= molecular pharming)

phase problem: In X-ray crystallography, the gap between the requirements for reconstruction of a model - for each reflection (defined by the angle of its diffraction), the amplitude (related to intensity) and phase (a measure of the alignment of the wave function with that of other reflections) - and the experimentally accessible quantities, the intensities of each reflection. (*see also* diffraction pattern; Fourier transformation)

phase transition: A melting-like phenomenon that occurs when lipid bilayers are warmed, usually below the naturally occurring temperature of a plasma membrane; due to thermal disruption of the regular stacking of fatty acid side chains. (*see also* transition temperature)

phenol-emulsion reassociation technique (PERT): A method to accelerate several thousand-fold the rate of DNA reassociation. An aqueous solution of single-stranded DNA with a chaotropic salt is shaken with phenol. When the resulting emulsion separates into two phases, the polynucleotide is found to have rapidly reassociated into double-stranded DNA. (*see also* FPERT)

phenotype: (*see* genotype)

pheromone: A chemical signal that travels through air or water from one organism to another.

phi-angle (angle): A characteristic of a peptide bond; the angle formed at the nitrogen atom by the bonds to the -carbon and to the carbonyl carbon of the neighbouring residue. (*see also* psi-angle (-angle); Ramachandran plot)

pH-jump: (*see* relaxation)

phorbol ester: A tetracyclic diterpene tumour-promoting substance originally derived from croton oil.

phosphate gripper: A basic oligopeptide sequence of many enzymes that protects their phosphate-containing substrate or transition state from interaction with water.

phosphoacylglycerol (phosphoglyceride): Phosphatidic acid or a derivative, e.g. phosphatidylcholine, phosphatidylserine.

phosphoanhydride: A structure formed by the removal of the elements of water from between two phosphoric acids, e.g. pyrophosphate, ADP.

phosphodiester: A diester of phosphoric acid; in nucleotide chemistry may be an internal diester, e.g. cyclic AMP, or an internucleotide bond as in RNA and DNA.

phosphogluconate oxidative pathway: (= pentose phosphate pathway)

phosphoglyceride: (= phosphoacylglycerol (phosphoglyceride))

phosphoinositide cascade: A series of events that is initiated extracellularly and leads via activation of phospholipase C to the liberation from membrane phospholipids of inositol 1,4,5-trisphosphate and diacylglycerol, which act intracellularly to increase cytosolic calcium and to activate protein kinase C respectively.

phospholipid: A covalent association of phosphoric acid and lipid, e.g. a phosphoacylglycerol, sphingomyelin.

phospholipid-transfer protein (exchange protein): A lipoprotein of the intracellular membranes of eukaryotic cells that can exchange its phospholipid or sterol for membrane components, and can shuttle lipids between these membranes, e.g. between the endoplasmic reticulum and Golgi apparatus.

phosphorescence: A phenomenon similar to fluorescence in which there is a long lag between excitation and emission. (*see also* triplet state)

phosphoroclastic split: A cleavage in which the elements of phosphate are inserted across a carbon-carbon bond, e.g. the phosphoroclastic reaction of pyruvate, in which pyruvate and orthophosphate are oxidized by a flavoprotein to acetylphosphate and CO_2.

phosphorolysis: Analogous to hydrolysis, the cleavage of a covalent bond by insertion across it of the elements of phosphoric acid, e.g. the glycogen phosphorylase reaction.

phosphorothioate: A phosphate analogue, especially a nucleotide analogue, in which a sulphur atom replaces an oxygen in one of the phosphate groups.

phosphorylation potential: A quantitative measure of the energy status of a cell: [ATP]/[ADP][P_i]. (*see also* adenylate energy charge)

photo-activated repair: A mechanism for repair of DNA by reduction of thymine dimers by DNA photolyase; contrasted with *dark repair*, an ATP-dependent process that involves excision of the dimers, resynthesis of the sequence by a polymerase and rejoining of the ends by a ligase.

photoaffinity labelling: A variant of affinity labelling in which the covalent attachment of a ligand to its site is effected by illumination of a light-sensitive chemical group on the ligand, often an azido group.

photocycle: The series of transformations undergone by rhodopsin from its capture of radiant energy, i.e. *cis-trans* isomerization of the all-*trans* form and regeneration of all-*trans*-rhodopsin.

photoisomerization: The light-induced rearrangement of one of the double bonds of all-*trans*-retinal.

photoluminescence: (*see* chemiluminescence)

photon: The fundamental particle of light.

photorespiration: The reaction of ribulose bisphosphate carboxylase in which molecular oxygen replaces carbon dioxide as a substrate, to form 3-phosphoglycerate and phosphoglycolate rather than two 3-phosphoglycerates.

photosynthesis: The light-dependent chlorophyll-catalysed biosynthesis by green plants of carbohydrate and oxygen from carbon dioxide and water. (*see also* Calvin cycle (reductive pentose cycle); C_3-C_4 photosynthesis; C_3 photosynthesis; C_4 photosynthesis (C_4 cycle; Hatch-Slack pathway); Hill reaction)

photosynthetic unit (PSU): The light-harvesting complex and reaction centre. Light is absorbed at 875 nm by proteins with caretenoid prosthetic groups, and the energy is transferred to photosystem II (PSII) of the reaction centre for the oxidization of water to oxygen and the transfer the electrons via cytochrome b6f to photosystem I (PSI) for the synthesis of NADPH. Especially at low light intensity, additional radiation may be collected by an antenna system, lower wavelength light-harvesting complexes, and transfered via the higher wavelength collectors to the reaction centre.

photosystem I (PSI): The light-driven reactions of photosynthesis tha abosrb at 700 nm or below and result in generation of NADPH. (*see also* photosystem II (PSII))

photosystem II (PSII): photosystem II PSII; the light-driven reactions of photosynthesis that abosrb at 680 nm and above and result in oxidation of water to molecular oxygen. (*see also* photosystem I (PSI); red drop)

phototroph: An organism that derives its energy by trapping radiant energy from the sun or other light source. (*see also* chemotroph)

phycobilin: (*see* bilin)

phycofluor: A cytological probe; a synthetic conjugate of an intensely fluorescent phycobiliprotein with a molecule like an antibody that gives it specificity.

phylogenetic tree (evolutionary tree): A reconstruction of the evolutionary relationship of contemporary species that shows their divergence from common ancestors and the branch points at which different species separated. A common ancestor is presented at the base of a central stem, and the tree ramifies as it ascends, often with distance from the base roughly related to evolutionary time. The relationships it displays are usually inferred from the similarities of amino acid sequences of contemporary proteins or of the base sequences of contemporary nucleic acids.

physical map: A goal of the human genome project; the characterization of each chromosome by landmarks, sequence-tagged sites and contigs, between which restriction maps may give finer detail.

physio-: A prefix derived from the Greek for nature.

phyto-: A prefix derived from the Greek for plant.

phytoextraction: (*see* phytoremediation)

phytofiltration: (*see* phytoremediation)

phytoremediation: The use of plants to detoxify soil or water. Phytoremediation may involve phytoextraction, the uptake of toxic chemicals by the roots of plants and their concentration in plant tissues, phytovolatilization, the chemical transformation of a toxic substance by a plant, usually by methylation, into a gaseous form, or phytofiltration, the use of crude plant extracts to selectively remove toxic materials from solutions.

phytovolatilization: (*see* phytoremediation)

Pi: Inorganic orthophosphate, i.e. PO_43^- and protonated forms.

pi electron-pi electron (electron- electron) interaction: A form of non-covalent interaction seen in proteins; the planes of two aromatic side chains overlap so that the orbitals can interact.

picket fence porphyrin compound: A model oxygen-binding porphyrin in which bulky side chains on one face of the ring permit binding of molecular oxygen while preventing oxidation of Fe^{2+}, which is co-ordinated on the other face of the ring by an organic base.

pico-: A prefix which denotes 10^{-12}; e.g. picomolar, 10^{-12} M.

piezophile: (*see* extremophile)

pigylated: (= glypiated)

ping pong: (*see* enzyme mechanism)

pinocytosis: (*see* endocytosis)

pitch: A feature of a helix, the distance parallel to the axis that corresponds to one turn of 3608, e.g. 5.4Å (0.54nm) for an -helix. (*see also* repeat; rise)

pKa: The negative of the common logarithm of the dissociation constant of an acid; i.e. $-\log K_a$, where $K_a=[H^+][A^-]/[HA]$ for $[HA]=[H^+]+[A^-]$.

plaque assay: A technique for quantification of infectious phage particles by counting plaque-forming units.

plaque hybridization: (*see* colony hybridization)

plaque purification: A technique to select a bacterial strain with a desired genetic trait; a mixture of cells is streaked on to a solid support and the resultant colonies are screened (e.g. for antibiotic resistance or a DNA sequence complementary to a specific probe), and the selected colony is then restreaked for further rounds of purification.

plaque: (*see* plaque-forming unit (PFU))

plaque-forming unit (PFU): A measure of viable bacteriophage particles, determined by plating a known volume of a solution on to a bacterial lawn and subsequently counting the areas of bacterial lysis (*plaques*).

plasma cell: A differentiated B-cell that secretes antibodies.

plasma membrane: The lipid bilayer and associated proteins and other molecules that surround a cell.

plasma: In biology, the liquid phase of whole blood. (*see also* serum)

plasma-desorption mass spectrometry: (*see* mass spectrometry (MS))

plasmalemma: (= plasma membrane)

plasmid: A self-replicating extra-chromosomal element, usually a small segment of duplex DNA that occurs in some bacteria; used as a vector for the introduction of new genes into bacteria.

plasmolysis: The dehydration of a cell caused by exposure to a hypertonic solution.

plasticity: The formation of new synapses in the mature brain, which is the presumed molecular and cellular basis for learning and memory.

plastid: A membrane-bound intracellular structure.

plate: A surface of solid growth media, often in a Petri dish, on which a bacterial culture can be grown; also the action of streaking a culture on the media.

platelet: A small membrane-limited fragment of cytoplasm derived from blebs from a megakaryocyte; platelets aggregate to form a plug during early stages of haemostasis.

pleated sheet: (= beta ()-pleated sheet)

plectonemic: Descriptive of the interaction of two DNA strands, either single- or double-stranded, in which an oligonucleotide of one strand is twisted into a plait of helices, one around the other, and requires the nicking of one strand to form or dissociate such an interaction. When one of the DNA strands is circular, the structure, a ring pierced by the linear strand, is called a *hemicatenane*. Plectonemic is contrasted with *paranemic*, descriptive of an interaction between DNA strands that is formed and dissociated without the breaking of covalent bonds and therefore constitutes a weaker interaction than a plectonemic joint. Plectonemic and paranemic are often used to describe joints formed during recombination events. A related folding is *toroidal*, in which a plait of helices is circularized and forms a ring, or *torus*.

pleio-: A prefix derived from the Greek for more, which refers to a multiplicity of causes or effects.

pleiotropic mutation: A mutation that affects expression of several characteristics, e.g. the effect of mutation of a promoter on structural genes regulated by it.

pleiotropism: A characteristic of action of some hormones; expression of several activities, e.g. the action of insulin both to promote glucose uptake by skeletal muscle cells and to promote protein synthesis.

pleuro-: A prefix derived from the Greek for the ribs or side, which denotes the membranes, pleura, that line the thorax and enclose the lungs.

pleuropneumonia-like organism: (= mycoplasma)

ploidy: In cytology and genetics, the number of chromosome complements of a cell, e.g. diploidy is the condition of two of each chromosome.

pluri-: A prefix derived from the Latin for several.

plus and minus method: A variant of the Sanger method of DNA sequencing. A primer is extended by a polymerase to generate a population of newly synthesized deoxyribonucleotides of assorted lengths; the unused dNTPs are removed, and polymerization continues in four pairs of plus and minus reaction mixtures; the minus mixtures have three NTPs and the plus mixtures have only one. After a second polymerization, the mixtures are fractionated by gel electrophoresis, and each plus and minus pair is compared to indicate the length of the new polydeoxyribonucleotide (by the mobilities of the bands) and the position at which polymerization had terminated as a result of the absence of the missing dNTP.

(+)-strand: (= plus-strand)

plus-strand: The coding DNA strand during transcription, as opposed to the non-coding *minus-strand.*

P nucleotide: (*see* V(D)J recombination)

pocket: An invagination of the surface of a protein where it can bind a ligand. (*see also* positive and negative selection (PNS))

poikilothermic: Descriptive of an organism that cannot maintain a fixed internal temperature. (*see also* homoiothermic)

point mutation: A change in a single nucleotide residue in a nucleic acid, which frequently causes a single amino acid residue replacement in the protein product of the gene.

polarization: A measure of the mobility of a fluorophore: $P=(I-I)/(I+I)$, where I is the intensity of emission and and indicate polarization that is parallel and perpendicular respectively to the exciting light. A *mobile fluorophore* is able to reorient itself within its fluorescence lifetime and therefore emits unpolarized light; an *immobile fluorophore* does not reorient and hence emits light polarized in the plane of excited light. (*see also* anisotropy)

pole of a cell: (*see* centriole)

poly-: A prefix that indicates many; a high degree of polymerization, as in polynucleotide, polypeptide. (*see also* mono-; oligo-)

poly(A) tail: The post-transcriptional modification of the 3'-end of eukaryotic mRNA; a sequence of 20-200 adenylates in 5'-to-3' linkage.

poly(A)- RNA: [*see* poly(A)$^+$ RNA]

poly(A)+ RNA: Polyadenylated RNA; usually synonymous with mature eukaryotic mRNA; contrasted with *poly(A)*$^-$ RNA, which is non-polyadenylated RNA.

polyamine: A compound with more than one amino group that often contains short chains of carbon atoms separated by a nitrogen atom and forms a secondary amine; associated in the cell with nucleic acids; e.g. spermine, spermidine, putrescine.

polyampholyte: A large molecule or small polymer that contains many acidic and basic groups. (*see also* isoelectric focusing (IEF); polyelectrolyte)

polycistronic mRNA: A single mRNA that contains the information necessary for the production of more than one polypeptide; characteristic of some prokaryotic mRNAs. (*see also* cistron; monocistronic mRNA; operon)

polyclonal antibody: A heterogeneous immunoglobulin preparation that contains antibodies directed against one or more determinants on an antigen; the product of daughters of several progenitor cells that have been programmed for immunoglobulin synthesis and secretion. (*see also* monoclonal antibody (mAb))

polydisperse: (*see* monodisperse)

polyelectrolyte: A compound or polymer with many charges of the same kind, e.g. polyglutamate, polylysine.

polyene: A multiply unsaturated compound, especially one that has strong visible absorbance bands and is capable of transducing light, e.g. chlorophyll, retinal.

polygeneous: (*see* heterogenous)

polyketide: A compound derived from biosynthetic acetyl and/or propionyl units and other building blocks, in which carbonyl carbons condense with a-carbons analogously to the synthesis of fatty acids. They are often fungal metabolites such as macrolides, which are formed by *polyketide synthases* (*PKSs*), modular multienzyme complexes encoded as clusters of open reading frames (ORFs) spanning more than 100kbp of the microbial genome, with each ORF responsible for a different enzyme activity. The cluster for any given PKS recruits the ORFs for those transferases that can form the carbon skeleton and the reductases and dehydratases that transform it into the particular polyketide product.

polyketide synthase: (*see* polyketide)

polylinker: A short DNA sequence that contains several restriction sites; intended to be placed into a vector so that the sites may subsequently be used for insertion of other DNA sequences.

polymerase chain reaction (PCR): A technique to amplify a specific region of double-stranded DNA. An excess of two *amplimers*, oligonucleotide primers complementary to two sequences that flank the region to be amplified, are annealed to denatured DNA and subsequently elongated, usually by a heat- stable DNA polymerase from *Thermus aquaticus* (*Taq* polymerase). Each cycle involves heating to denature double-stranded DNA and cooling to allow annealing of excess primer to template and elongation of the primers by the *Taq* polymerase; the number of *amplicons*, i.e. the target sequence fragments between flanking primers, doubles with each cycle. (*see* self-sustained sequence replication; strand displacement amplification)

polymorphic amplifiable restriction endonuclease fragment (PARF): A natural polymorphism detectable with probes constructed to a representation of the affected parent DNA.

polymorphism: Appearance in several forms, e.g. of a protein that is subject to tissue-specific post-translational modification, or of mRNA that is processed through alternative splice sites; also the appearance in a population of more than one structural gene at a single genetic locus.

polymorphonuclear leucocyte (PMN): One category of blood cell; a neutrophil.

polyol: A simple sugar whose carbonyl function has been reduced, e.g. sorbitol, mannitol.

polyprotein: A primary product of protein synthesis, a single polypeptide chain that will eventually be cleaved into several separately functional proteins; e.g. the precursor produced by retroviruses that gives rise to several proteins, among them a coat protein, a polymerase and an endopeptidase.

polysome display: An alternative to phage display as a method for selection of a protein with optimized binding properties. A cell-free bacterial expression system transcribes a DNA library and the resulting mRNAs are translated into proteins. The protein and the mRNA are prevented from dissociating from the polysomes, so when the complexes are selected for a protein which binds to an immobilized ligand, the polysome and mRNA are also selected. The mRNA is isolated and reverse transcribed and amplified by PCR to enter another cycle of transcription, translation and selection. (*see* in vitro evolution and ribosome display)

polysome: A structure that is functional in protein synthesis, formed by several ribosomes attached to a single mRNA molecule.

polytene chromosome: An amplified form of an insect chromosome in which the chromatin is packed in irregularly spaced transverse bands.

polytopic: Descriptive of a protein that spans a membrane at least once and has domains that project into both cytoplasmic and extracellular spaces.

pool: The aggregate of metabolites that are in a functional equilibrium in a cell or an organism, e.g. cytosolic hexose phosphates, mitochondrial tricarboxylic acid cycle intermediates, liver glycogen. (*see also* compartment)

porcine: Derived from the pig.

porin: A bacterial membrane channel that allows free transport of small hydrophilic substances.

porter: A protein or group of proteins in a membrane that is responsible for facilitated transport.

position effect: The dependence of transgene expression on its site of incorporation into the host genome.

positional candidate approach: A strategy for identification and cloning of a new gene; a combination of *functional cloning*, in which a property of the protein product (e.g. immunogenicity or liganding) is used to scan a cDNA library, and *positional cloning*, a more laborious approach that relies on chromosome walking from a nearby known sequence identified by genetic linkage.

positional cloning (reverse genetics): The search for a gene (e.g. that for cystic fibrosis) with no knowledge of the gene product by locating it on the basis of linkage analysis with other genetic traits or individual polymorphisms in non-coding regions, followed by chromosome walking to identify candidate open reading frames that can be tested by expression in appropriate tissues or by zoo blotting, according to knowledge of the phenotype. Alternatively, the search for the gene for a specific protein from knowledge of its amino acid sequence (functional cloning), such that the sequence of part of the gene is deduced from the amino acid sequence and a labelled oligonucleotide probe complementary to it is synthesized. The probe allows one to screen appropriate DNA libraries for the gene. (*see also* positional candidate approach)

positive control: The activation of transcription in a bacterium in response to its metabolic state, e.g. synthesis of several enzymes in response to glucose deprivation. (*see also* glucose effect)

positive co-operativity: A form of allosteric behaviour in multimeric enzymes in which binding of substrate to one subunit increases the affinity of the other subunits for the substrate. (*see also* negative co-operativity)

positive energy balance: The dietary situation in which caloric dietary intake is greater than energy expenditure, resulting in growth and possibly obesity. (*see also* negative energy balance)

positive-inside rule: The generalization that, for proteins that span a membrane, the region that extends into the cytoplasmic space is positively charged.

post-labelling: A method for detection of DNA modification. After exposure of DNA to a xenobiotic, it is digested with nucleases and labelled with ^{32}P; the resulting mixture is then separated by any of several conventional techniques to detect non-natural labelled fragments.

post-synaptic: Descriptive of a nerve cell that receives its signal from another nerve cell across a synapse. (*see also* pre-synaptic)

post-transcriptional modification: The processing of RNA subsequent to its synthesis. This which may include hydrolysis and transesterification of phosphodiester bonds and modification of bases. For eukaryotic mRNA it includes capping of the 5'-end, polyadenylation of the 3'-end and splicing of introns.

post-translational modification: The processing of a protein subsequent to its synthesis; includes phosphorylation, glycosylation, acetylation, limited proteolysis, attachment of prosthetic groups and cross-linking. (*see also* co-translational)

potentiometric titration: The use of a potentiometer, especially a pH meter, to follow the titration of an oxidizable group with a reductant (or vice versa) to generate a graph of reduction potential against percentage reduction, or the titration of an acid with a base (or vice versa) to generate a graph of pH against amount of titrant.

potocytosis: (*see* caveolae)

potonuon: (*see* junk DNA)

Potter-Elvehjem homogenizer: A device used to disrupt tissues. A cylindrical glass or hard polymer pestle rotates in a close-fitting tube and a suspension of the tissue particles is subjected to shearing forces as the pestle moves up and down and presses the suspension through the space between the rotating pestle and the tube.

PPi: Pyrophosphate; P_2O74^- and protonated forms.

pre-: A prefix derived from the Latin for before, and signalling precendence in time , place, sequence, etc.

preassembled signalling complex: The assembly on the surface of a cell, especially at caveolae, of receptor and signal transduction proteins (e.g. G-proteins, tyrosine kinases).

prebiotic: Descriptive of the synthetic reactions that occurred in geological time before the advent of life.

precipitin reaction: The cross-linking of antigens through bivalent antibodies that creates an insoluble three-dimensional matrix.

precursor protein: (*see* protein splicing)

precursor: An antecedent metabolite whose structure is modified by one or more enzymic reactions of a metabolic pathway.

pre-edited region: (*see* guide RNA (gRNA))

pre-incubation: The equilibration of some components before initiation of a reaction (the incubation) by addition of an essential substrate, cofactor or enzyme.

pre-initiation complex: In replication, transcription or translation, a complex that is potentially composed of template, enzymes and initiation factors but that is incomplete for optimal function due to the absence of one or more elements. (*see also* initiation complex)

pre-mRNA: One class of heterogeneous nuclear RNA; DNA transcripts that will (usually) be modified by polyadenylation and capping and processed to remove introns before being used in translation. (*see* pre-mRNA splicing, spliceosome)

pre-mRNA (heterogeneous nuclear mRNA): The RNA that is the direct transcript from DNA and therefore includes introns before it is processed into mature mRNA.

pre-mRNA splicing: The process by which an intron is excised and exons are religated in the post-transcriptional modification of RNA. The process is effected in vivo by a complex, called a spliceosome, of small nuclear RNA-proteins (snrps) and other proteins that assemble on a pre-mRNA and by hybridizing with the pre-mRNA at splice site consensus sequences, force the appropriate regions of the pre-mRNA into a reactive configuration and catalyse the excision of an intron in a process mechanistically similar to group II self-splicing. Snrps act by a two-step mechanism that cleaves the phosphodiester bond between the intron and the upstream exon, leaving a 3'-hydroxyl group; the 5'-phosphate of the intron is transferred to the 2'-hydroxyl usually of an adenine nucleotide (the branch point) of the intron near its 5' limit; the 3'-hydroxyl of the upstream exon then displaces the 3'-terminal nucleotide of the intron from the downstream exon, thus ligating the two exons and leaving the intron as a lariat, a rho-shaped structure that with the 5'- terminus anchored to the branch point. *In vivo*, the process is catalysed by large nuclear ribonucleoprotein particles (lnRNPs), 200 S particles (500 Å, approx. 500 KDa), composed of four subunits, each of which appears to perform like a snrp.

prenyl: A chemical group composed of multiples of the branched, unsaturated, 5-carbon isoprene unit, e.g. geranyl (10 carbons), farnesyl (15 carbons), geranylgeranyl (20 carbons). It is sometimes attached to the thiol group of an O-methylated C-terminal cysteine residue that forms part of a *CAAX box*, where C is the cysteine, A is any amino acid with an aliphatic side chain and X is any amino acid; a prenylpyrophosphate donates the prenyl group to the -CAAX sequence, AAX is proteolytically cleaved and the exposed carboxy group of the derivatized cysteine residue is enzymically methylated by *S*-adenosylmethionine. Prenylated proteins serve in the control of cellular function. Vitamins A and K, and dolichol pyrophosphate, carry reduced prenyl groups.

preparative in situ hybridization (prep-ISH): A coincidence cloning technique for isolation of DNA fragments that are characteristic of a small, well characterized portion of a chromosome. Restriction-endonuclease-digested DNA is annealed to segments of chromosomes that are fixed to a solid support and then washed to remove non-successfully hybridized fragments; the hybridized fragments are eluted and PCR-amplified. (*see also* coincidence painting (chromosome painting))

preparative ultracentrifugation: The separation of subcellular organelles by high-speed centrifugation.

prep-ISH: (= preparative in situ hybridization (prep-ISH))

preproprotein: The entire polypeptide that is encoded by an mRNA for a secretory protein before processing by proteinases cleaves the signal sequence and another polypeptide sequences that must be removed before the protein becomes functional. The sequence is preproproteinproprotein (inactive) protein (active). (*see also* prohormone; zymogen)

pre-synaptic: Descriptive of a nerve cell that transmits its signal to another nerve cell across a synapse. (*see also* post-synaptic)

prey: (*See* two-hybrid system)

pri-: A prefix derived from the Latin for first.

Pribnow box: (= TATA box)

primary structure: The amino acid sequence of a protein or polypeptide, or the nucleotide sequence of a polynucleotide. (*see also* quaternary structure; secondary structure; supersecondary structure; tertiary structure)

primer: An RNA sequence hybridized to a DNA template whose elongation by a DNA polymerase constitutes DNA synthesis. A *random primer* is a mixture of polynucleotides with all four bases at each sequence position; an *arbitrary primer* is a single species with a single base at each sequence position.

primer extension analysis: A method for measurement of the amount of a specific mRNA and its length relative to a known sequence. A labelled synthetic primer designed for complementarity to a sequence in the mRNA, usually close to the 3'-end, is reverse transcribed to generate a cDNA that by the intensity of its labelling indicates the amount of mRNA originally present, and by its mobility on gel electrophoresis indicates its length, and hence the length of the mRNA from the site of primer attachment to its 5'-end.

primer mismatch analysis: A method for detection of the presence of a specific DNA fragment. Amplification by PCR with, alternatively, a wild-type-specific or a mutant-specific primer for the same chain and an anti-sense chain primer generates an amplification product only where there is a perfect match of primers to target.

primer walking: A method for sequencing relatively long pieces of DNA by use of a library of short, and therefore inexpensive, polynucleotides to assemble a primer complementary to a known sequence, initially a vector sequence. Three or four appropriate hexanucleotides are assembled on a 18- or 24-base sequence of the template with suppression of the binding of individual hexanucleotides by competition from a single-stranded DNA-binding protein. The downstream sequence is determined as far as is practical, and a newly identified sequence is used to select and bind three or four new hexanucleotide primers for identification of the subsequent downstream sequence.

primosome: A complex of proteins that first synthesizes the RNA primer for DNA replication that is subsequently elongated to form an Okazaki fragment. (*see also* transcriptional activation)

prion: A proteinaceous infectious particle which causes one of a number of non-inflammatory, slowly-developing degenerative neurological diseases which are characterized by extracellular plaques, notably bovine spongiform encephalopathy (mad cow disease), scrapie in sheep and Creuzfeld-Jakob disease in humans. Unlike other infections agents, it is apparently composed entirely of a single protein, PrP^{Sc}, which differs from the normal cellular protein, PrP^{C}, only in its conformation; the benign conformation is converted into the infectious conformation when the two come together, by a rare spontaneous conversion or by a mutation of PrP which allows it to more easily assume the infectious conformation.

pro-: A prefix derived from the Greek for before, which indicates priority, e.g. in time or causality.

proband: (= propositus (proband))

probe: (= hybridization probe)

processing: Post-transcriptional modification of RNA or post-translational modification of a polypeptide or protein.

processivity: The property of an enzyme that acts repetitively on a single macromolecule, such as a polymerase on its template or a kinesin on myosin, to remain attached to the polymer between cycles of action.

prochirality: The symmetry characteristic of a molecule, or of a centre within a molecule, by which like groups may be distinguished, e.g. the property of citric acid that allows aconitase to distinguish between its two carboxymethyl groups. (*see also* chirality; CIP classification; meso-carbon; R; symmetry)

procollagen suicide: The magnified effect of a mutation in the propeptide of one collagen precursor molecule on the structure of secreted collagen. The phenomenon results from the disruption of fibril assembly when a small proportion of the subunits can form abnormal cross-links, e.g. disulphide bridges.

proenzyme: (= zymogen)

progenote: In early evolution, an organism with a rudimentary mechanism for replication and translation. (*see also* protogenote)

progestogen: A compound, usually a steroid, that prepares the uterus for implantation and supports pregnancy, e.g. progesterone.

programmed cell death (PCD): (*see* apoptosis)

prohormone: An inactive precursor of a hormone, especially a polypeptide hormone.

prokaryote: (*see* eukaryote)

promoter: A sequence of dsDNA that regulates the binding and activity of RNA polymerase. The core promoter includes sequences within 40 nucleotides of a transcription start site; e.g. the initiator, TATA box, downstream promoter element, which separately or together direct synthesis by RNA polymerase II. Further upstream from the initiation site is the promoter-proximal region, which contains additional regulatory sequences. The core promoter and the promoter-proximal region are, collectively, the promoter. (*see also* insulator)

proneurophysin: (*see* neurophysin)

proofreading (editing): In DNA replication, the 3'-to-5' exonuclease activity of a polymerase that removes incorrectly paired bases, also referred to as editing; in protein synthesis, the hydolysis of an incorrectly paired amino acyl-tRNA by the amino acyl-tRNA synthetase. (*see also* excision; kinetic proofreading)

proofwinding: Analogous to proofreading, a putative activity that tests the accuracy of a transcription by hybridization of the product RNA to an authentic complementary polynucleotide.

propensity: The likelihood of finding an amino acid residue in a certain secondary structure, e.g. glycine in an -helix. Data from statistical surveys, host-guest analysis, model peptides, directed mutagenesis or molecular dynamics calculations are expressed as *G* and quantified in kJ/mol (kcal/mol).

propeptide: The part of a protein that is proteolytically cleaved during the protein's maturation, e.g. the internal residues of proinsulin, the N-terminal residues of pepsinogen. The term is usually not used for a signal peptides, but for the peptides removed from protein precursors that can be isolated and that have a reasonably long half-life.

propinquity effect: (*see* entropy effect)

propionate rule: Analogous to the acetate and isoprene rules, the observation that some natural products are assembled from multiple propionate (propionyl-CoA and methylmalonyl-CoA) units in head-to-tail condensations.

propositus (proband): In human genetics, the individual whose forebears and descendants are under investigation.

proprotein: A precursor protein that requires post-translational modification for its activity to be expressed, e.g. pepsinogen, trypsinogen. (*see also* preproprotein)

prosegment: (*see* zymogen)

prosome: (= proteasome)

prostaglandin: An eicosanoid that features a cyclopentane nucleus.

prosthetic group: A moiety that is tightly attached to a protein, often a participant in an enzymic reaction; e.g. haem, FAD, Zn^{2+}, pyridoxal phosphate. (*see also* apoprotein; holoprotein)

protamine: A small, helical, arginine-rich basic protein that lies along the major groove of B-DNA.

protease: (= endopeptidase)

proteasome: A multicatalytic proteinase; a molecular organelle of size up to 26 S (2000 KDa) in eukaryotes. The basic 20 S (700 KDa) structure is a barrel-like structure composed of two inner rings, each of 7 -type catalytic subunits and two outer rings, each of 7 -type subunits. Protein substrates enter the narrow opening of a subunit ring and are cleaved into oligopeptides (hepta- to nona-peptides) as they pass through the inner opening of the subunit ring. This is the initial phase, the 'bite', of a proposed 'bite-chew' mechanism. Subsequent proteolytic steps, the 'chew', degrade the initial oligopepitde products to amino acids. On the outer surface of the ring of -type subunits may be located additional subunits that confer ATP-dependence of proteolysis and greater specificity, e.g. for polyubiquitinated proteins.

protection footprinting: (*see* footprinting)

protein: A linear polymer of a-amino acids held together by peptide bonds. A protein may include other components attached covalently (e.g. carbohydrate, phosphate, fatty acid, biotin) or non-covalently (e.g. Zn^{2+}, FAD, haem), and may be cross-linked by disulphide bridges or, less commonly, by other specialized bonds. It may exist as a complex of two or more identical or dissimilar polypeptide chains.

protein cage: The reaction site within which inorganic materials are deposited and that can reach the supramolecular (i.e. nanometer) size, e.g. apoferritin, within which hydrated ferric oxide is deposited to form ferritin.

protein class: A classification according to the content of -helix and -pleated sheet. A protein is designated if it has a minimum number of residues in an -helix and fewer than a maximum in -pleated sheets (suggested to be 45% and 5%, respectively); it is , if it has less than a maximum of -helix and more than a minimum of -sheet (5% and 40%, respectively); +, if it has minimum of both -helix and anti-parallel -sheet (15% and 15%); and /, if it has a minimum of both -helix and parallel -sheet. It is irregular () if it has less than 5% -helix and 10% -sheet.

protein kinase: An enzyme that uses ATP to phosphorylate a group on a protein, e.g. a serine, threonine or tyrosine hydroxy group. (*see also* MAP kinase; protein kinase C (PKC); tyrosine kinase)

protein kinase C (PKC): A calcium-dependent protein kinase; often one stimulated by diacylglycerol.

protein kinesis: The directed cellular movement of proteins, e.g. nuclear proteins to the nucleus, lysosomal proteins to the lysosomes, etc.

protein ladder sequencing: Determination of a protein's amino acid sequence by the generation of a series of fragments from the original protein which represent sequences of single amino acid removals. The series is generated by stepwise degradation effected by Edman degradation using phenylisothiocyanate and a small amount of a termination reagent, phenylisocyanate. Cycles of reaction are performed without isolation of intermediates, and the products are characterized by mass spectrometry.

protein nucleic acid (PNA): A polyamide backbone, e.g. (-NH-CH_2-CH_2-NR-CH_2-CO-$)_n$, which bears purine and pyrimidine bases (R in the formulation) and can mimic oligonucleotides by hybridizing with ssDNA or RNA. Some PNAs can even form triplexes with DNA strands, and can invade a dsDNA strand to dissociate the base-pairing of the DNA and form a duplex with one strand. PNAs are uncharged, and therefore form more stable complexes with a single stranded nucleic acid, especially at low ionic strength, than the corresponding double-stranded nucleic acid.

protein processing: Limited proteolysis that converts a biologically inactive translation product into an active enzyme, hormone, structural protein, etc.

protein quality: The relative dietary value of a protein that reflects its content of the various essential amino acids.

protein ruler: A high-molecular-mass protein that functions in organization of cytoskeletal structures of uniform length by serving as a template for aggregation of exact numbers of other components of the structure along its length, e.g. nebulin, which is suggested to align the subunits of skeletal muscle thin filaments.

protein splicing: The excision of an intervening protein sequence, an *intein*, from a precursor protein, with peptide bond formation between the flanking domains (the *N-extein*, originally attached to the N-terminal end of the intein, and the *C-extein*, originally attached to its C-terminal end) to produce the spliced protein. Intein sequences are homologous with homing endonucleases that transfer the intron gene to an intron-less allele, and some free intein proteins possess the endonuclease activity. The intein-C-extein interface is an asparagine-serine bond. The splicing mechanism is autocatalytic, and is proposed to employ displacement of the N-extein from the intein by the -hydroxy group of the serine. This produces a branched intermediate: to the C-extein serine residue are attached the intein in a peptide bond to the -amino group and the N-extein in an ester bond to the -hydroxy group. The asparagine residue cyclizes to free the intein, with a C-terminal aminosuccinimide residue formed from the asparagine; the N- and C-exteins remain joined by an ester bond. Displacement of the N-extein from the -hydroxy group by serine's

-amino group produces the spliced protein. Some C-exteins have threonine or cysteine in place of the serine.

protein truncation test (PTT): *In vitro* synthesized protein assay; a procedure for detection of mutations that result in a protein of significantly different size, e.g. by early termination, insertion, change in reading frame or altered mRNA splice site. cDNA is prepared from RNA of a selected cell or tissue; PCR is performed using as a primer an oligonucleotide containing the promoter of the desired RNA polymerase and eukaryotic translation initiation sequence. The resulting amplicon can be translated so as to include a label (e.g. a radioactive amino acid or biotin-lysyl-tRNA) which will permit visualization of the protein when separated by SDS/PAGE.

proteinase: (= endopeptidase)

protein-associated thermosynthesis (PAS): (*see* thermogenesis model)

proteodermochondran sulphate: A polymer of the extracellular matrix, derived from chondroitin 4-sulphate that is incompletely epimerized and sulphated. The completely processed polymer is dermatan sulphate. (*see also* glycosaminoglycan; proteoglycan (mucopolysaccharide))

proteoglycan (mucopolysaccharide): A complex composed of glycosaminoglycans that radiate from a protein core in a bottlebrush-like structure. There are two major types of membrane proteoglycans: *glypicans*, which are anchored to the peripheral surface by a glycosylphosphatidylinositol linkage; and *syndecans*, which contain transmembrane regions and short cytoplasmic tails. (*see also* core protein; link protein)

proteolipid: A protein that is soluble in organic solvents, usually due to the presence of a significant number of fatty acids esterified to secondary hydroxy groups and/or fatty acylation of its N-terminus.

proteome: The repertoire of proteins that exist in a cell or tissue under a particular set of conditions

proteomics: The application of techniques of protein analysis, especially highly automated robotics, to the design and discovery of new drugs and diagnostic methods.

proteonomics: (*see* omic research)

protist: A unicellular eukaryote.

proto-: Borrowing from historical linguistics, a prefix that denotes the simplest possible precursor of two or more organisms, inferred from phylogenetic trees reconstructed from polynucleotide sequences, e.g. protogenote (but not proto-oncogene).

protofilament: The linear subunit of a microtubule; composed of alternating - and -tubulin subunits. Thirteen protofilaments arranged into a cylinder constitute a microtubule.

protogenote: In early evolution, the most recent common ancestor of cellular organisms (which include the archaebacteria, eubacteria and eukaryotes) with only rudimentary translation capabilities. (*see also* progenote)

protome: The basic subunit of a multimeric protein.

proton inventory: (*see* solvent isotope effect)

proton motive force: The free energy, expressed in volts, represented by a pH gradient across the membrane of a vesicle due to the concentration gradient plus the membrane potential. This energy is used by coupling the transport of electrons across the membrane with the synthesis of ATP. (*see also* chemiosmotic theory)

proton motive Q cycle: The cyclic reduction and oxidation of cytochromes *b* and c_1 and an iron-sulphur protein, which comprise the cytochrome bc_1 complex. In each cycle one ubiquinol is oxidized to ubiquinone, two cytochrome c_{ox} are reduced to cytochrome c_{red} and four protons (two oxidized from ubiquinol and two from water) are transported from the inner, electronegative, surface to the outer, electropositive, surface of mitochondria and of many bacteria.

proton pump: A mechanism for the active transport of a proton across a membrane, e.g. the proton pump of the mitochondrial inner membrane, of the gastric parietal cell, of the thylakoid membrane.

proton switch: In enzyme chemistry, a mechanism of tautomerization in which a proton is removed by a solvent molecule from an ionizable intermediate in the catalytic cycle and added to another atom of the intermediate from another solvent molecule.

protonophore: (*see* ionophore)

proto-oncogene: A normal cellular gene that has the potential to become an oncogene by mutation or translocation; identical with c-oncogene. (*see also* anti-oncogene (tumour suppressor gene))

protoplast: A plant cell that has been stripped of its cell wall so as to be limited only by its plasma membrane. (*see also* spheroplast)

provirus: A stage in the development of a virus, in which it is incorporated into the genome of a host cell; in a retrovirus, the double-stranded DNA stage.

proximity effect: (*see* entropy effect)

pseudo-: A prefix derived from the Greek for false, which indicates something related but inauthentic, incomplete or inactive.

pseudo-first order: In kinetics, descriptive of a rate that is directly dependent upon the concentration of one reactant that is experimentally varied. The rate may also be dependent upon the fixed concentrations of other reactants.

pseudogene: A DNA sequence that is homologous to a structural gene, but cannot be expressed because it has no continuous open reading frame. It often occurs without introns.

pseudoglobulin: (*see also* globulin)

pseudo-intron: A sequence of an RNA transcript that sometimes behaves like an intron and is spliced out, but sometimes remains in the mature mRNA. (*see also* alternative splicing; read-through)

pseudo-knot RNA: (*see* ribozyme)

pseudomutant: The product of a silent mutation.

pseudopod: A rounded projection of an animal cell used for locomotion. The cytoplasm streams into the pseudopod and extends the cell in its direction.

psi: A symbol that in a designation of an amino acid sequence denotes a structure that substitutes for a peptide bond, e.g. Gly-(CH_2-CO)-Gly is H_2N-CH_2-CH_2-CO-CH_2-COOH.

psi-angle (-angle): A characteristic of a peptide bond; the angle formed at the a-carbon by the bonds to the nitrogen atom and the carbonyl carbon atom. (*see* phi-angle (angle); Ramachandran plot)

psychrophile: (*see* extremophile)

puckering: The divergence from planarity of a non-aromatic ring compound, e.g. a sugar or steroid, to accommodate the bond angles of the constituent atoms. (*see also* chair form; half-chair conformation)

puff: A local expansion of a chromosome due to the unfolding of DNA as it is being transcribed.

puffer fish: (*see* Fugu rubripes)

pulse labelling: In metabolic studies, the exposure of a biological system (e.g. a tissue slice, a cell culture, mitochondria) to a radiolabelled metabolite over a relatively short time, in order to subsequently follow the passage of the label from precursor metabolite to product metabolite. A variant is *pulse-chase labelling*, in which the pulse is terminated by the addition of a large amount of the unlabelled metabolite in order to rapidly lower its specific radioactivity. (*see also* equilibrium labelling)

pulse-chase experiment: (*see* pulse labelling)

pulsed amperimetric detection (PAD): A method for detection of chromatographic eluates, especially oligosaccharides, not easily detected by other means; a cycle of less than 1s first applies a voltage across the eluate, adsorbs the carbohydrate to an electrode and oxidizes it (the oxidizing current is what is monitored), followed by a strong oxidizing voltage which cleans the electrode for the start of the next cycle.

pulsed-field gel electrophoresis (PFGE): A technique for separation of chromosomes and very large DNA molecules, usually for purposes of genetic mapping; alternating pulses of electricity directed at 1208 across the plane of an agarose gel slab drive the structures through the gel.

pump: An energy-dependent mechanism by which a metabolite or ion is forced across a membrane against a concentration gradient.

punctuation: In translation, the polynucleotide sequence that signals the initiation or termination of a message.

purine: One of the two classes of heterocyclic organic bases that are found in nucleic acids (principally adenine and guanine) and in several other kinds of biological compounds, e.g. coenzymes, nucleotides, sugar derivatives; also includes methylated purine alkaloids such as caffeine and theobromine. (*see also* pyrimidine)

purple membrane protein: A constituent of *Halobacterium halobium*; a retinal-containing protein that traps light and transports protons across the membrane in which it is embedded.

putrefaction: The process of decay of organic material by bacterial, fungal or non-biological processes into foul-smelling products.

pyr-: A prefix derived from the Greek for fire, which indicates a product of heating.

pyranose: The form of a sugar when it is condensed into a six-membered ring, which consists of five carbon atoms and the oxygen atom that is the link to the anomeric carbon. (*see also* furanose)

pyridine nucleotide: NAD^+, NADH, $NADP^+$ or NADPH.

pyrimidine dimer: A covalent attachment of two adjacent bases that can form when ultraviolet light damages double-stranded DNA; e.g. thymidine dimer.

pyrimidine: One of the two classes of heterocyclic organic bases that are found in nucleic acids (principally uracil and cytosine in RNA; uracil and thymine in DNA) and in several other kinds of biological compounds, e.g. nucleotides, sugar and lipid derivatives. (*see also* purine)

Q

q: In genetics, a designation of the short arm of a chromosome. (*see also* centromere)

Q_{10}: The factor by which a reaction is accelerated upon raising the temperature by 10 degrees. (*see also* activation energy)

QC-PCR: Quantitative competitive PCR. (= quantitative PCR (QPCR))

Q_{CO2}: The rate of production of CO_2 by a preparation, expressed as l/h per tissue dry weight. (*see also* manometry)

Q_{O2}: The rate of uptake of O_2 by a preparation, expressed as l/h per tissue dry weight. (*see also* manometry)

QPCR: (= quantitative PCR (QPCR))

Q replicase: A viral RNA-dependent RNA polymerase; used to greatly amplify concentrations of an RNA that serves as a template.

quad: (*see* enzyme mechanism)

quadrapole mass filter: The instrumentation for mass spectroscopy that allows selection of ions of a specific m/e^- ratio. As the desorbed material travels down an electric field toward the detector, it passes through a space boarded by four parallel rods, which are electrodes. All four electrodes have the same DC current, onto which is superimposed an AC current that oscillates at a rate calculated to retain within the four rods only ions of the desired m/e^- ratio and to deflect all other ions. Thus, the signal-to-noise ratio is kept high. By varying the charges on the electrodes, the spectrum of ions can be sampled. An alternative design is the ion-trap, in which the desorbed ions enter a cavity bounded front and back by circular entrance and exit endcap electrodes, and on the sides by a ring electrode. The voltages on all three electrodes are varied so that in-coming ions, passing through aperture in the entrance endcap electrode are directed into a circular orbit inside the ring electrode, until a sufficient number are accumulated, at which time, they are directed through an aperture in the exit endcap electrode toward the detector in a sequence that reflects their m/e^- ratio.

quadruplex DNA: A form of DNA in which four oligo(G) sequences, either of the same or of different strands, line up either in a parallel, an anti-parallel or in a fold-back (mixed parallel and antiparallel) pattern. The interior of quadruple the helix has four guanine bases in Hoogstein base pairings; each plane of four guanines is separated from the adjacent plane by a Na^+ or K^+ ion. Although these structures have not been identified cells, they are strongly suspected to cap the ends of chromosomal DNA. *see* telomere

quantitative PCR (QPCR): A method for quantification of a polynucleotide by inclusion with it of a known amount of an easily distinguished template as an internal standard to compensate for variation in efficiency of amplification. At the termination of cycling, the relative amounts of the products are measured by ethidium bromide binding and fluorescence or by incorporation of ^{32}P-labelled primers.

quantitative competitive PCR: (= quantitative PCR (QPCR))

quantum yield: The ratio of the number of molecules that respond, e.g. by reacting or fluorescing, to the number of photons absorbed.

quaternary structure: The arrangement in space of polypeptide subunits that make up a multimeric protein. (*see also* primary structure; secondary structure; supersecondary structure; tertiary structure)

quenching: In fluorescence spectroscopy, a decrease in emission due to a variety of causes that dissipate energy of the excited state, e.g. transfer of energy to a non-fluorescent molecule such as molecular oxygen.

quiescent: Non-proliferating, as applied to cells.

quiescent affinity label: A reagent that reacts specifically with the active site of the enzyme for which it was designed. It bears a generally reactive chemical group, such as a chloromethane, and substrate-like moieties that permit it to bind only to the target enzyme. (*see also* affinity labelling)

quinimine form: A tautomer of the Schiff base adduct of pyridoxal phosphate with an amino acid in which the pyridine ring nitrogen is protonated but uncharged and the carbon atom of the coenzyme that attaches the amino group of the amino acid has lost a proton. It is an intermediate between the *aldimine form* (in which the pyridine ring is protonated and charged and the attachment of the amino acid to the cofactor is a Schiff base) and the *ketimine form* (which can best be represented as a Schiff base formed by condensation of an a-oxo acid with pyridoxamine).

quinoprotein: A protein, usually of microbial origin, with a quinone prosthetic group: pyrroloquinoline quinone (e.g. methanol dehydrogenase), 6-hydrodopaquinone (topa quinone, e.g. amine oxidase) or 2',4-bitryptophane-6,7-dione (tryptophan tryptophanylquinone, e.g. amine dehydrogenase).

quinozyme: (= quinoprotein)

quorum-sensing: The ability of some micro-organisms to monitor cell density cues, such as acylhomoserine lactones, and to adjust their gene expression accordingly.

R

R: A designation of absolute configuration at an asymmetrical centre; classified as *R* (*rectus*) or *S* (*sinister*) according to the arrangement in space of its substituents, which are assigned precedence according to sequence rules. Highest priority is given to the highest atomic number. If for an asymmetrical carbon atom, when viewed from the direction opposite the substituent of lowest priority, the sequence from the highest to lowest priority of the remaining substituents follows a clockwise path, the configuration is *R*; if counter-clockwise it is *S*. (*see also* chirality; CIP classification; prochirality)

R band: (*see* AT queue)

R loop: (*see* AT queue)

R strain: (*see* lipopolysaccharide (LPS))

RACE: (= rapid amplification of cDNA ends (RACE))

RACE cleavage: (= RecA-assisted restriction cleavage by endonuclease (RACE cleavage))

racemic mixture: The mixture in equal amounts of two enantiomers, e.g. D- and L-alanine.

rack model: (= induced-fit theory (rack model))

radiation-induced hybrid mapping: A method for location of a human genetic marker within a restricted region of a single chromosome. A hybrid cell line that contains a specific single human chromosome in each recipient rodent cell is heavily irradiated, and surviving cells are selected that carry fragments of the human chromosome integrated into the rodent chromosome. Each rodent chromosome is subsequently characterized by its own morphology and genetic markers and by the human fragment it carries, so that the cells may be analysed, e.g. by fluorescence *in situ* hybridization, for a new genetic marker.

radiationless transfer: The fall to the ground state of an excited molecule without the emission of light, i.e. without fluorescence, so that the energy appears as heat.

radioautography: (= autoradiography)

radioimmunoassay (RIA): A competitive binding assay that depends upon displacement of a radiolabelled ligand from an antibody by the non-labelled standard or test sample, followed by separation of the labelled ligand into bound and unbound fractions. (*see also* ELISA)

raft: A cluster of sphingolipids and cholesterol, with associated glycophosphoinositol-anchored and/or signal transduction proteins, in a lipid bilayer membrane. (*see* caveolae)

Ramachandran plot: A graph that displays calculated energy isotherms for conformations of peptide backbones; *phi*-angles against *psi*-angles; energy minima correspond to -structure, -helix and the collagen triple helix.

Raman optical activity: (*see* vibrational optical activity (VOA))

Raman scattering: (*see* light scattering)

random: (*see* enzyme mechanism)

random activation of gene expression (RAGE): A procedure for creating a genome-wide library of cDNAs by use of several insertion vectors, each containing a promoter sequence linked to an exon and followed by an unpaired splice site. Transfection into eukaryotic cells allows the vectors to randomly insert by non-homologous recombination. Where the insertion is upstream from a host cell gene, it activates expression of a transcript that encodes a chimeric protein that containing the vector-encoded sequence and some or all of the host cell protein.

random coil: The unfolded, denatured, state of a protein or nucleic acid.

random DNA amplification (RDA): A method to produce improved strains, e.g. salt tolerance of a plant, by gene amplification.

Random genome fragments are introduced into a vector and grown in a bacterium. The plasmid is then introduced into the original strain at the DNA sequence homologous with the DNA fragment, under conditions that favor tandem repeats. Modified strains are selected first by a resistance marker that was part of the original plasmid, then by cycles of selection under pressure of the desired property, e.g. at high salinity.

random oligonucleotide mutagenesis: (= saturation mutagenesis (random oligonucleotide mutagenesis))

random peptide library : For purposes of identification of polypeptides that have a high affinity for a target structure, the assortment of, e.g., the 520 to 720 penta- to hepta-peptides that may be incorporated into a phage genome and expressed on the phage surface. (*see* phage display)

random primer: (*see* primer)

random walk: The irregular path followed by a particle that shows Brownian movement or by a flagellar bacterium that periodically changes its direction of motion.

randomization: The migration of an isotope from one position in a labelled metabolite to another position due to the conversion of the metabolite into a symmetrical compound, in which labelled and non-labelled positions become equivalent, or by a process in which moieties of a metabolite are transformed into one another; e.g. the fate of the carboxy and methyl carbons of acetate as it passes through the reactions of the tricarboxylic acid cycle.

rapid amplification of cDNA ends (RACE): A PCR-based method to obtain cDNA clones of rare transcripts which begin with only a partial internal sequence. The 3'- and 5'-ends are analysed separately; a single (-)-strand is first constructed from the 3'-end of the transcript by use of reverse transcriptase and a poly(dT) primer. The (+)-3'-end is then generated in multiple copies by PCR using poly(dA) and the known internal partial sequence as primers. A single (-)-strand copy is constructed from the 5'-end using reverse transcriptase with a (-)-primer based on the known internal sequence. This 5'-(-)-transcript is tailed with poly(dA) and then used as a template in a PCR with the known internal sequence and poly(dA) as primers. Once the 3'- and 5'-ends are known, primers can be constructed to produce multiple copies of the gene by PCR. (*see also* RecA-assisted restriction cleavage by endonuclease (RACE cleavage))

rapid mixing: (*see* stopped-flow)

RAP-PCR: RNA fingerprinting arbitrary primer PCR. (*see* differential PCR display)

rare-cutter: A restriction nuclease that is specific for nucleotide sequences that occur infrequently in the genome.

ras gene: An oncogene of the murine sarcoma virus. (*see also* ras protein)

ras protein: A product of the *ras* gene; a G-protein.

rate-limiting step: The slowest step of a metabolic pathway or enzymic reaction; the one that determines the rate of appearance of the ultimate product.

Rayleigh scattering: (*see* light scattering)

RCS: Recognition consensus sequence. (*see* shape library)

RDA: (= recommended daily allowance (RDA), or representational difference analysis (RDA))

rDNA: Recombinant DNA.

reaction centre: In photosynthesis, the energy-transducing unit. By absorption of light it excites a chlorophyll dimer, which can then transfer a pair of electrons across a membrane to a quinone, which will accept two protons from its microenvironment to generate a hydroquinone.

reaction co-ordinate: A graphical representation of the course of a reaction that shows the energy of the reactants and products as the reaction progresses.

reactive immunization: A technique to widen the scope of catalytic reactions effected by abzymes. Reactive compounds are used as antigens; some of the resulting antibodies are directed to the reaction complex with a protein or carbohydrate. The antibody can then recruit to its binding site the components of the reaction and force them into reactive juxtaposition.

reactive nitrogen species (RNS): Radicals such as NO· (nitric oxygen radical) and NO^+ (nitrosyl) which are chemically reactive, especially with proteins or nucleic acids.

reading frame: The register in which the translation apparatus senses the information coded within an mRNA molecule. As the code is a triplet, there are three possible reading frames. (*see also* open reading frame (ORF))

read-through: A protein that is the product of alternative splicing of the transcription product. A potential intron remains in the mRNA and is translated rather than excised. (*see also* pseudo-intron)

real time quantitative PCR : A method for quantification of nucleic acid sequences by PCR amplification in the presence of doubly-labelled probe. The 5'-end contains a reporter fluorophore, e.g. carboxyfluorescein, and the 3'-end a group which can quench the fluorescence of the first, e.g. carboxytetramethylrhodamine. As PCR products are generated, the probe hybridizes to them and the 5'-nucleotidase activity of the *Taq* polymerase removes the reporter group; as it diffuses away from the quenching group, its fluorescence emission becomes a direct measure of the extent to which the amplification has progressed. As fluorescence may be determined in the same wells of microtiter plates in which the PCR reaction is performed, so the fluorescence determination requires no additional handling and is in 'real time'. An internal standard, e.g. the gene for -actin or a housekeeping enzyme, allows normalization and interpretation of the data. The RT-PCR variation, in which reverse transcriptase is first used, allows quantification of RNA.

RecA-assisted restriction cleavage by endonuclease (RACE cleavage): A method for cleavage of double-stranded DNA at specific restriction sites. A suitable restriction site is selected, around which the base sequence is known, and an oligonucleotide, up to 50 or 60 bases in length, is synthesized to complement it. It is hybridized to the DNA to form a triple helix, which is stabilized with the help of the RecA protein. Methylation then blocks all but the protected restriction site, so that subsequent treatment with the restriction nuclease, in the absence of the protecting oligonucleotide and RecA protein, will cleave only at the designated site. The use of two protecting oligonucleotides for sites that span a sequence of interest allows excision of a specific fragment from a DNA preparation as complex as the human genome.

receptor: The binding site for a hormone or neurotransmitter that initiates its action at the cellular level. First proposed about a century ago to fulfill the postulate that agents may act only if they can bind to a target, receptors are now recognized chemical structures.

receptosome (endosome): A coated vesicle from which the clathrin shell has been removed.

recessive: (*see* allele)

recoding: The altered reading of an mRNA due to instructions (recoding signals) embedded in the sequence, which cause frame shifts or other reading alterations.

recognition consensus sequence: (*see* shape library)

recombinant DNA: DNA formed by bringing together DNA fragments from different species.

recombinant PCR: A protocol for site-directed mutation and construction of a vector containing the mutation. A pair of homologous complementary primers that contain the mutation are used separately with other primers each of which, with its mutagenic primer, bracket a unique restriction site in order to construct two double-stranded oligonucleotides that overlap only over the sequence of the mutagenic primers. These double-stranded DNAs are mixed, denatured and annealed to form some of a hybrid that is double-stranded only in the sequence of the mutagenic primers and single-stranded in the direction that extends towards the other external primer sites. PCR amplification and restriction fills in the full-length DNA and prepares it for insertion into a vector.

recombination: The natural or synthetic production of a new DNA molecule from polynucleotide sequences that originate from more than one parent DNA molecule.

recombination joint: The site of interaction between two double-stranded DNA molecules; at regions where the two double strands are homologous, equivalent strands of each duplex break and exchange complementary strands to form a bridge between the two DNAs. (*see also* Holliday model)

recombination signal sequence: *see* V(D)J recombination

recommended daily allowance (RDA): The intake of nutrients, vitamins, minerals and protein considered advisable by the Ministry of Agriculture, Fisheries and Food (U.K.), or the Food and Nutrition Board of the National Academy of Sciences or the Food and Drug Administration (U.S.A.); usually greater than the minimum daily requirement.

red drop: The sharp decline in photosynthetic efficiency as the wavelength of exciting light enters the red region (above 680nm) due to the insensitivity of photosystem II to excitation by red light.

red muscle (slow-twitch muscle): A well vascularized form of skeletal muscle with many mitochondria and much cellular myoglobin; supplied with energy mainly by -oxidation. (*see also* white muscle (fast-twitch muscle))

redox: Descriptive of a chemical or biochemical reaction that involves oxidation of one reactant or substrate at the expense of reduction of another.

redox potential (Eo'): The electrical potential, expressed in volts, of a reductive half-reaction, e.g. for $NAD^{+}+2e^{-}+H^{+}=NADH$, *E*8'=-0.32V.

reducing sugar: A sugar whose anomeric carbon atom can be oxidized by an alkaline cupric ion solution, e.g. Fehling's solution, Benedict's solution. (*see also* non-reducing sugar)

reduction: In its essential form, the addition of electrons to a compound; possibly with the addition or abstraction of other atoms.

reduction division: The first phase of meiosis, in which a diploid cell gives rise to two haploid daughter cells.

reduction potential: (= redox potential (Eo'))

reductive pentose cycle: (= Calvin cycle (reductive pentose cycle))

REF: (= restriction endonculease fingerprinting (REF))

reflectional symmetry: (*see* symmetry)

refractive index: The ratio of the velocity of light in a vacuum to the velocity in the medium inquestion.

regio-: A prefix derived from the Latin for district or location, which in chemistry refers to a site within a molecule.

regioselective: (*see* regiospecific)

regiospecific: Descriptive of a reaction that proceeds by a unique course even though more than one course is formally possible, e.g. an addition that occurs at only one atom of an unsymmetrical olefin; contrasted with regioselective, in which the reaction shows merely a preference for one course over another; and with stereospecific, in which the reaction produces one diastereoisomer rather than another.

regulability: (*see* regulation)

regulation: The phenomenon of adjustment of metabolite flux, e.g. glucose that enters the anaerobic glycolysis pathway and exits as lactate, whatever its molecular mechanism. It reflects for each step of the pathway the range over which that step's activity can be modulated (regulability) and the effect this has on overall metabolic flux (control). Flux-control, metabolite-control, elasticity and response coefficients have been used to model and analyse the effects of substrate, inhibitor and allosteric effector concentrations.

regulatory subunit: A part of a multimeric enzyme; a protein that binds to and inhibits the catalytic subunit unless it is bound to a regulating molecule; e.g. in the case of cyclic AMP-dependent protein kinase, the subunit that, unless bound to the cyclic nucleotide, inhibits the catalytic subunit.

regulon: A group of genes the expression of which is co-ordinated, e.g. by the actions of the same transcription factors.

reiteration: Repetition of identical, or nearly identical, long DNA sequences in a chromosome.

relative contact number: (*see* contact number)

relaxation: In kinetics and spectroscopy, the return to equilibrium of a macromolecule after a very brief perturbation of the environment, e.g. recovery of a protein from a sudden burst of energy that allows a fast transient increase in temperature (temperature-jump) or a fast transient change of pH (pH-jump).

relaxed: Descriptive of the topology of closed circular DNA with no superhelical torsion applied to it, which has one turn per 10.4 base pairs.

relaxed bacteria: (*see* magic spot nucleotides)

relaxed conformation: R-form. (*see* Monod-Wyman-Changeux (MWC) model)

release factor: A protein that binds to the terminator codon of mRNA and allows hydrolysis of the peptidyl-tRNA ester bond.

releasing factor: (= releasing hormone)

releasing hormone: A hormone, often of hypothalamic origin, whose function is to cause release of another hormone, often of hypophyseal origin; e.g. thyrotropin-releasing hormone.

renaturation: The return to native structure from a denatured state; in nucleic acid chemistry, identical to annealing.

reni-: A prefix derived from the Latin for kidney.

renin: An aspartate proteinase that is secreted by the kidneys and releases angiotensin I from angiotensinogen in the blood; not to be confused with rennin.

rennin: Obsolete name for chymosin, a milk-clotting aspartate proteinase secreted by the stomachs of newborn mammals; not to be confused with renin.

repeat: A feature of a helix; the distance parallel to the axis in which the backbone structure repeats itself. For a DNA double helix, for example, the repeat distance is 34Å (0.34nm), which represents 10 residues. (*see* also pitch; rise)

reperfusion injury: The damage caused by upon resumption of blood flow following ischaemia. The mechanism may involve the mitochondrial permeability transition.

repertoire: In molecular immunology, the assortment of immunoglobulins produced by an animal; this assortment changes (repertoire shift) as the animal develops immunity following a challenge.

repertoire cloning: A method to produce human monoclonal antibodies. mRNAs are isolated from a tissue that generates immunoglobulins (e.g. spleen) and a reverse transcriptase makes cDNA copies of them. The Fd regions, i.e. the H-chain component of the Fab fragment (the VH and CH1 domains of the immunoglobulins regions), are selected and amplified by PCR to obtain a heavy-chain library. Similarly, a light-chain library is obtained. The two populations of cDNA are combined into a random pairing to form combinatorial constructs, inserted into coliphage vector along with a promoter, and used to infect E. coli. Plaques are then screened for ability to bind an antigen. (*see also* synthetic peptide combinatorial library (SPCL))

repertoire shift: (*see* repertoire)

repetitive DNA: Sequences of DNA that are found to be repeated, sometimes thousands of times over.

repli-: A prefix derived from the Latin for reply.

replication: Synthesis of new DNA strands complementary to existing template strands.

replication fork. The site where the two polynucleotides of a parent DNA separate during replication. The daughter strands are attached to the two separated polynucleotides that trail away from the fork as it advances into the parent DNA. (*see also* Okazaki fragment)

replication licencing factor: (*see* licencing)

replicon: A segment of the eukaryotic genome that contains several genes and is replicated as a unit from a single origin. Replication is bi-directional; its boundaries are the points where replication from one origin meets the replication fork which advances from the opposite direction. Each replicon is presumed to be identical to a DNA loop.

reporter group: A chromophore or fluorophore that is sensitive to its environment; a group added to a protein to indicate changes in its environment as the protein is manipulated, e.g. by denaturation or association with a ligand or other proteins.

representation: A portion of the genome with lowered complexity (e.g. by cleavage with an infrequent cutting restriction nuclease, ligated to oligonucleotide adapters so it can then be amplified by PCR), but that can be used in subtractive cloning and kinetic selection procedures. (*see also* representational difference analysis (RDA))

representational difference analysis (RDA): A method for isolation and amplification of DNA sequences that differ between two populations (a subtractive technique); e.g. changes associated with antigenic variation in micro-organisms, activation of the immune system, chromosomal translocation. The complexity of one population, the tester DNA, is lowered by cleavage with an infrequent cutting endonuclease, ligated to adaptor oligonucleotides to allow PCR amplification; the other population, the driver DNA, is similarly treated, but after amplification the ligated adaptors are removed to prevent amplification. These amplicons do not include the entire parent population, but serve as representations. Tester and an excess of driver DNA are mixed, denatured and annealed. By using kinetic enrichment conditions, the tester strands common to the two populations will be annealed to driver strands; only sequences unique to the tester population will form duplexes. The ends of the duplexes are filled in with a polymerase and PCR amplified; only fragments found uniquely in the tester population will amplify exponentially, as hybrid duplexes undergo only linear amplification and therefore will be a small fraction of the amplified fragments.

repressor: A protein that can regulate transcription by binding to the operator and causing repression.

RER: (= rough endoplasmic reticulum (RER))

rescue: A strategy in chemotherapy in which lethal doses of an antifolate are given to block nucleotide biosynthesis in rapidly proliferating cells, followed by treatment with 5-formyltetrahydrofolate or thymidine to 'rescue' normal cells.

residualizing label: A moiety attached to a circulating protein in order to subsequently detect by histological means the site of the protein's degradation in a multicellular organism. A reporter group, such as a radioactive or fluorescent group that does not alter the protein's turnover rate, is attached to the protein under non-denaturing conditions, and is retained in lysosomes of those cells that degrade the protein.

residue: The part of a single sugar that appears in a polysaccharide; of a single amino acid in a protein; of a single nucleotide in a nucleic acid, etc.; usually the monomer minus the elements of water.

resistance transfer factor (R-factor): A plasmid that contains genes for resistance to several antibiotics, which permits the transfer of drug resistance between bacteria.

resolution: In X-ray crystallography, the precision with which atoms are located in space; usually expressed in Å (10-10m).

resonance energy transfer: A phenomenon in which one fluorophore excites another when there is significant overlap of the emission spectrum of the primary fluorophore with the excitation spectrum of the secondary fluorophore and some other conditions are met. In part it depends strongly upon the distance that separates the fluorophores and their orientation toward one-another; it is useful in calculation of intermolecular distances of macromolecules.(*see also* molecular ruler)

resonance scattering: (= anomalous scattering (resonance scattering))

resorption: Reabsorption; the reversal of an excretion, e.g. re-uptake of Na+ in kidney tubules, or of a deposition, e.g. degradation of bone mineral by osteoclasts.

respiration: The consumption of oxygen by an organism, organ, tissue, cell or subcellular structure, as a terminal electron acceptor in the conversion of metabolites into carbon dioxide.

respiration state: The condition of tightly coupled mitochondria that is characterized by their respiratory rate (all but state 3 are slow) and defined by the rate-limiting component: state 1, ADP- and substrate-limiting; state 2, substrate-limiting; state 3, respiratory chain-limiting; state 4, ADP-limiting; state 5, oxygen-limiting.

respiratory burst: The sudden consumption of oxygen by a leucocyte when it is stimulated.

respiratory chain: (= electron transport chain (respiratory chain))

respiratory control index: A quantitative measure of the degree of coupling of phosphorylation to oxidation; the ratio of the rates of respiratory state 3 to state 4.

respiratory quotient (RQ): The molecular ratio of carbon dioxide generated to oxygen consumed.

resting metabolic rate (RMR): The rate of oxygen consumption by a person at rest which represents the energy required for essential functions such as respiration, circulation, maintenance of temperature. Also known as the basal metabolic rate (BMR).

resting potential: The membrane potential of an excitable cell in the absence of stimulation.

restriction endonculease fingerprinting (REF): A method for analysis of large restriction fragments of DNA with a high degree of confidence of finding mutations. The large fragment is separately digested with several cocktails (preferably five) of endonucleases. The resulting digests are denatured, mixed together, labelled and electrophoresed on non-denaturing gels. In the restriction component of the analysis, additional or absent electrophoretic bands, compared with normal DNA, will result when a mutation creates, destroys or changes the electrophoretic mobility of a band.

restriction endonuclease: One of a group of enzymes that cleave internal phosphodiester bonds of both strands of DNA at specific nucleotide sequences, especially at palindromic sequences. (*see also* isoschizomers)

restriction fragment length polymorphism (RFLP): The variation of size of polynulceotides which are generated by restriction endonucleases and visualized by electrophoresis and Southern blotting. Changes in the normal pattern are indicative of major deletions or insertions in the observed genetic marker.

restriction fragment: A fragment of double-stranded DNA that results from the cleavage of DNA by a restriction endonuclease.

restriction map: The characterization of double-stranded DNA by the locations of the sites at which various restriction endonucleases can cleave it. (*see also* contig map; linkage map; physical map; sequence-tagged site (STS))

restriction-site-generating PCR (RG-PCR): A method for detection of a specific mutation that does not itself cause a unique restriction fragment length polymorphism, because it neither creates nor removes a known restriction site. A new restriction site is created by the introduction of a deliberately mismatched PCR primer, the 5'-end of which is designed to hybridize to the mutant template and the 3'-end of which encodes a new restriction site. (*see also* mismatched PCR)

restrictive temperature: For a heat-sensitive cell line, a temperature at which it cannot function.

retentate: Non-filtered particles and macromolecules.

retention of configuration: The lack of a change in the configuration at an asymmetrical centre over the course of a reaction, e.g. the enzymic addition of an -glucosyl unit to glycogen from UDP-glucose, which is itself an -glucoside. Such reactions may involve two Walden inversions.

reticulo-: A prefix derived from the Latin for net, which indicates origin in a tissue of histologically net-like appearance.

reticulocyte: A non-nucleated (in mammals) precursor of an erythrocyte, still capable of haemoglobin synthesis

reticuloendothelial system: The tissues involved in synthesis and degradation of blood cells and their constituents; includes Kupffer, bone marrow and spleen cells.

retinoid: A sesquiterpene related to or derived from retinol.

retroelement: A mobile genetic element; either a retrovirus or a retrotransposon, both of which are characterized by long terminal repeats. (*see also* long interspersed elements (LINES); retrotransposon)

retrograde: Having a direction opposed to the normal or usual, e.g. transport from a synapse to the cell body of a neuron or from the distal to the proximal cisternae of the Golgi apparatus; as distinguished from anterograde, having the normal or usual direction.

retro-inverso: Descriptive of a polypeptide analogue in which the direction has been inverted internally, giving it two N- or two C-termini, e.g. ...CO-CHRNH-CO-NHCHR'CO...

retrotransposon: A mobile genetic element composed of repetitive DNA sequences that flank open reading frames, including one that codes for a reverse transcriptase. In contrast to transposons, which in prokaryotes are transmitted as DNA, retrotransposons are transposed by reverse transcription of the RNA into cDNA, its conversion into double-stranded DNA and insertion into a new position in the genome. (*see also* retroelement)

retrovirus: *see* RNA virus)

reverse genetics: (= positional cloning (reverse genetics))

reverse physiology: (*see* orphan receptor)

reverse transcriptase: A DNA polymerase that uses an RNA template; an RNA-dependent DNA polymerase.

reverse transcriptase-PCR (RT-PCR): Also known as RNA phenotyping, RNA-PCR or message amplification phenotyping; a method for amplification of a specific mRNA by prior use of reverse transcriptase to form a cDNA, then use of PCR to amplify. (*see also* competitive PCR)

reverse turn: A secondary structure of proteins in which the polypeptide backbone turns sharply on itself in as few as three residues, and is stabilized by hydrogen bonds between the amide hydrogens and carbonyl oxygens on either side of the turn. Also known as a -bend, a -turn or a hairpin bend.

R-factor: (= resistance transfer factor (R-factor))

R-form: Relaxed conformation. (*see* Monod-Wyman-Changeux (MWC) model)

RF: In thin layer and paper chromatography, the ratio of the distance moved by a test sample to the distance moved by the solvent front or, in gel electrophoresis, to the distance moved by a low-molecular-mass marker.

RFLP: (= restriction fragment length polymorphism (RFLP))

RG-PCR: (= restriction-site-generating PCR (RG-PCR))

rho protein: *see* anti-terminator; termination factor)

rhodo-: A prefix derived from the Greek for rose.

RIA: (= radioimmunoassay (RIA))

ribophorin: A protein of the rough endoplasmic reticulum that recognizes leader sequences of growing polypeptides and, through them, attaches the ribosomes.

riboside: Ribonucleoside. *see* nucleoside)

Ribosomal RNA: (= rRNA)

ribosomal stuttering: A phenomenon in which translation, e.g. of a retroviral polyprotein message, compensates for a frame shift in the message itself.

ribosome: A non-membrane-bound organelle; a complex of RNA molecules and proteins that is a site of protein synthesis in eukaryotes. (*see also* heavy subunit; light subunit)

ribosome display: An alternative to phage display as a method for selection of a protein with optimized binding properties, which is based on the principles of polysome display. A cell-free expression system transcribes a DNA library and the resulting mRNAs are translated into proteins. The protein and the mRNA are prevented from dissociating from the ribosome, so when the complexes are selected for a protein which binds to an immobilized ligand, the ribosome and mRNA are also selected. The mRNA is isolated and reverse transcribed to enter another cycle of transcription, translation and selection. *see* in vitro evolution)

ribosome jumping: 5'-Cap-dependent initiation of translation of some viral mRNAs in which a ribosome subunit recognizes the 5'-cap of the mRNA but 'jumps' over the 5'-non-translated region of strong secondary structure to reach the AUG start codon. *see* internal ribosome entry site; (see tripartite leader)

ribosome shunting: (= ribosome jumping)

ribosome-inactivating protein (RIP): One of a class of plant proteins that bind to elongation factor 2 and thus inactivate ribosomes.

Type 1 (e.g. the toxin gelonin) consist of single polypeptide chain proteins; type 2 (e.g. ricin) consist of two proteins linked by a disulphide bond, one the toxin and the other a lectin that attaches to recognition sites on a target cell.

ribotide: Ribonucleotide. *see* nucleotide)

ribozyme: An RNA molecule that owing to peculiarities in its folding, is able to catalyse the interchange of some of its phosphodiester linkages to achieve intramolecular splicing. Examples of ribozyme groups are hammerhead, hairpin and pseudo-knot RNAs, which are whimsically named for their topologies. Ribozymes normally cleave cis, i.e. intramolecularly, but can be engineered to cleave a designated RNA target trans, i.e. intermolecularly. (*see also* intein; self-splicing)

right-handed helix: By analogy with the right hand, a helix, e.g. of a polypeptide, that advances towards the C-terminus in the direction of the extended thumb as the backbone of a chain turns in the direction of the fingers closing on the palm. A left-handed helix has the opposite sense.

RIP: (= ribosome-inactivating protein (RIP))

rise: A feature of a helix. The distance parallel to the axis corresponds to one residue, e.g. 1.5Å (0.15nm) for an -helix. (*see also* pitch; repeat)

RLF: Replication licencing factor. (*see* licencing)

RL-RT-PCR: (= RNA ligase reverse transcription PCR (RL-RT-PCR))

RMR: (= resting metabolic rate (RMR))

RNA: Ribonucleic acid; a macromolecule formed of repeating riboses linked by phosphodiester bonds between the 3-hydroxyl group of one and the 5-hydroxyl group of the next. A purine, adenine or guanine, or a pyrimidine, cytidine or uracil, is held in a glycosidic bond to the anomeric carbon of the sugar. RNA has several biological functions, most of which depend upon its ability to form sequence-specific interactions with DNA. RNA comprises the genome of some viruses, and in some cases can perform some elementary catalytic functions. (*see also* mRNA; rRNA; tRNA)

RNA chaperone: A putative RNA-binding protein that assists RNA in attainment of its biologically active conformation. Non-specific RNA-binding proteins are proposed to prevent RNA becoming temporarily trapped in innumerable non-functional metastable conformations (*kinetic trapping*); specific RNA-binding proteins are proposed to stabilize the biologically active conformation. (*see also* chaperone machine)

RNA editing: (*see* guide RNA (gRNA))

RNA fingerprinting arbitrary primer-PCR: (*see* differential PCR display)

RNA interference (RNAi): The phenomenon of gene silencing effected by the interaction of a double-stranded RNA, approx. 22 bp long, with the cognate genomic DNA sequence. The RNA-DNA complex is subject to hydrolysis by a specific deoxyribonuclease.

RNA ligase reverse transcription PCR (RL-RT-PCR): A procedure for amplification of minute amounts of an unknown RNA (or DNA) sequence. The RNA fragment is ligated at both ends to single-stranded palindromic oligodeoxynucleotide that provides a site for subsequent hybridization with a second oligodeoxynucleotide. Reverse transcription uses as a primer the hybridized oligodeoxynucleotide, and then PCR amplification generates a family of polynucleotides that can be treated with a restriction endonuclease to produce large fragments of the cloned RNA and smaller fragments of linker oligonucleotides.

RNA recoding: Non-classical reading of the genetic code, e.g. frame-shifting during transcription to find a codon; use of AUG, normally a stop signal, as a coding signal.

RNA regulator: An mRNA-like transcript that is not translated but that affects expression of sequence-unrelated DNA.

RNA single-strand conformational polymorphism (rSSCP): A variant of the SSCP technique for detection of mutations in a DNA fragment in which mobility shifts of bands of mutant compared with normal DNA are more pronounced. A phage promoter is added to one of the PCR primers to allow production of an RNA transcript that is then analysed for altered electrophoretic bands on non-denaturing gel electrophoresis

RNA virus: A virus with an RNA genome that may be either an mRNA, (+)-RNA, or its complement, (-)-RNA. Class 1 contains (+)-RNA; class 2, (-)-RNA, which is the template for an RNA-dependent RNA polymerase; class 3, double-stranded RNA, in which (+)-RNA is synthesized by an RNA-dependent RNA polymerase; class 4, retrovirus, in which (+)-RNA is a template for an RNA-dependent DNA polymerase (a reverse transcriptase).

RNA world: In one scenario for the origin of life, a stage before the evolution of DNA and proteins in which replicative and catalytic functions were carried out exclusively by RNA molecules.

RNase cleavage: A method for detection of single base substitutions in DNA fragments. A labelled RNA probe that spans the target DNA sequence becomes susceptible to RNase A cleavage wherever there is any mismatch in sequences. The cleavage of the probe is detected upon gel electrophoresis.

RNP: RNA-protein complex. (*see* small nuclear RNA (snRNA))

ROA: Raman optical activity. (*see* vibrational optical activity (VOA))

Robison ester: Obsolete name for glucose 6-phosphate.

rocket electrophoresis: A technique for identification and quantification of a material by electrophoresis into a slab of agarose gel that contains an antibody. The height of the leading edge of the area of precipitation is indicative of the amount of antigen present.

rolling circle replication: Mechanisms of DNA replication of small circular genomes, such as some viruses and plasmids, in which the template generates multiple tandem copies. One strand is nicked and serves as a primer for elongation at the 3'-end, while the unnicked strand serves as a template. (*see also* rolling transcription)

rolling transcription: The synthesis of RNA from a circular DNA template. Some circular DNAs, even without a promoter sequence and without assistance of a primer, may be transcribed to form concatenated circular RNA. If the circular DNA encodes a ribozyme, the concatenates may process themselves to form monomeric RNA. (*see also* rolling circle replication)

Root effect: A phenomenon in some teleost fish that allows them to generate O_2 gas in their swim bladders in the face of considerable hydrostatic pressure. The haemoglobin structure and the local secretion of lactic acid create an exaggerated Bohr effect.

Rossmann fold: A supersecondary structure that is found in some proteins; a motif of -strands connected by -helix cross-over elements. (*see also* nucleotide-binding domain).

rotamer distribution: *see* entropy effect)

rotational catalysis: A proposed partial mechanism of ATP synthase in which a transmembrane ion channel, which is a hexamer composed of a ring of alternating and subunits, in which a single subunit rotates, binding ADP and Pi and displacing ATP.

rotational symmetry: *see* symmetry)

rough endoplasmic reticulum (RER): The endoplasmic reticulum that is studded with ribosomes; the site of synthesis of secretory, intrinsic membrane, Golgi apparatus and ER-resident lysosomal proteins.

RQ: (= respiratory quotient (RQ))

rRNA: Ribosomal RNA; the RNA that forms part of the structure of ribosomes.

rSSCP: (= RNA single-strand conformational polymorphism (rSSCP))

RT-PCR: (= reverse transcriptase-PCR (RT-PCR))

rubi-: A prefix derived from the Latin for red.

ruffled edge: (= lamellipodium (ruffled edge))

run-off: *see* nuclear run-off assay)

run-on assay: A method for evaluation of the status of replication, transcription or translation in a cell. The assay continues the synthesis of a polymer without reinitiation. (*see also* nuclear run-off assay).

S

-some: A suffix with the same derivation as soma-.

3'-splice site: (= acceptor (3') splice site)

5'-splice site: (= donor (5') splice site)

S: (*see* R)

S phase: (*see* cell cycle)

S strain: (*see* lipopolysaccharide (LPS))

S1 nuclease mapping: A technique for identification of nucleotides at the 5'-end of an mRNA which, because it retains at that end its original triphosphate moiety, may be specifically labelled for sequencing by the Maxam-Gilbert technique. A restriction fragment that is complementary to the 5'-end of the mRNA protects it against S1 nuclease digestion, which is specific for single-stranded RNA and DNA; the only remaining polynucleotides therefore are the 5'-end of the mRNA and its unlabelled complement.

SAAB: Selected and amplified (protein) binding site oligonucleotide. (*see* cyclic amplification and selection of targets (CASTing))

saccharide: A sugar or a polymer of sugars linked by glycosidic (acetal or ketal) bonds. (*see also* mono-; oligo-; poly-)

Sakaguchi reaction: A colorimetric reaction for identification and quantification of guanidino groups that involves reaction with -naphthol and sodium hypochlorite.

Salmonella test: (= Ames test (Salmonella test))

salt bridge: An interaction between positive and negative charges on side chains of a protein.

salting in: (*see* salting out)

salting out: The addition of a salt, especially ammonium sulphate, to decrease the solubility of a susceptible protein. Some proteins respond by becoming more soluble upon addition of a salt, i.e. salting in. (*see also* globulin)

salvage pathway: A metabolic reaction that recovers some of a partially degraded compound for resynthesis, e.g. resynthesis of GTP from guanine recovered from degradation of nucleotides via the hypoxanthine-guanine phosphoribosyltransferase reaction.

sandwich immunoassay: A method for quantification of an antigen large enough to have two epitopes. The antigen serves as a bridge between an immobilized (captive) antibody attached to one epitope, and a radioisotope-, fluorophore- or chromogenic enzyme-labelled antibody attached to a second epitope. The amount of immobilized label is directly related to the amount of antigen present.

Sanger method (dideoxynucleotide sequencing): A technique for determination of the base sequence of a homogeneous polynucleotide that acts as a template in a replication system. A small amount of a specific dideoxynucleoside triphosphate is included with the four normal deoxynucleotide triphosphates during replication. Chain termination occurs whenever a dideoxynucleotide is incorporated that generates a labelled polynucleotide whose chain length, evaluated by polyacrylamide-gel electrophoresis, is indicative of the distance from the 5'-end of the primer to the position at which the normal nucleotide occurs. Detection of the electrophoresis bands may be by a label such as 32P or, more recently, by a fluorophore incorporated into the primer (dye primer sequencing) or the dideoxynucleotide (dye terminator sequencing). In fact, in a variation of the dye terminator method, the labelling of each of the four dideoxynucleotides with a fluorophore that emits at a different wavelength permits the identification of the base that terminates each electrophoresis band within a separated mixture of all four (single lane sequencing). (*see also* Maxam-Gilbert method (chemical cleavage method); plus and minus method)

saponification number: A characteristic of an acylglycerol; the number of mg of KOH necessary to neutralize the fatty acids liberated from 1g of material.

saponin: A naturally occurring derivative of a plant steroid bound to a sugar moiety. Those with certain pharmacological properties, e.g. digitonin, are called cardiac glycosides.

SAR: [= scaffold-associated region (*see* AT queue), or structure-activity relationship (SAR)]

sarco-: A prefix derived from the Greek for flesh.

sarcolemma: The excitable plasma membrane of a muscle cell.

sarcoma: A malignant tumour of mesodermal origin.

sarcomere: The minimal functional unit of skeletal muscle; roughly a cylinder bounded on its ends by Z-lines where the thin filaments are anchored; these extend towards the centre. Thick filaments extend from the centre of the sarcomere, forming the A-region. The thick and thin filaments overlap in two darker-staining zones, which bracket the central H-region.

sarcoplasm: The cytoplasm of a skeletal muscle cell.

sarcoplasmic reticulum: A flattened membrane-limited compartment that surrounds a myofibril of skeletal muscle and that contains Ca2+, which can be released to stimulate contraction.

sarcosome: A mitochondrion of a muscle cell.

satellite DNA: Short repetitive DNA sequences that occur mainly at the ends or in the centre of chromosomes and are therefore suspected of serving structural roles; also, polynucleotides that are separable, on the basis of their characteristic density, from the bulk of nuclear DNA, and that have repetitive sequences.

saturation mutagenesis (random oligonucleotide mutagenesis): A technique to mutate all bases of a gene. DNA synthesis proceeds from an assortment of oligonucleotide primers synthesized by substitution at each point of the sequence of all three incorrect nucleotides, so that an unbiased assortment of mutants is generated for the span of DNA covered by the primer. Use of other modified primers to cover the length of the gene produces a complete library of mutants.

saturation: In protein chemistry, the limit at which a reversible binding site is fully occupied by a ligand. In magnetic resonance, saturation is the raising of a paramagnetic compound or chemical group to an activated state by radiofrequency radiation faster than it can spontaneously relax to the ground state; as the high-energy spin state of the molecules is fully populated, higher intensity radiation cannot be absorbed.

scaffold protein: One of the non-enzymic components of a multiprotein signal transduction complex. The complex is assembled upon activation of the receptor or is assembled but dormant until receptor activation. The scaffold is the framework for assembly at the cytoplasmic domain of a receptor; with the assistance of anchoring proteins it recruits kinases, phosphatases and other enzymes, and, with the assistance of adaptor proteins, other factors which will continue the signal sequence within the cell.

scaffold: Protein engineers use the term to refer to a domain or small protein that is the object of mutation intended to introduce or refine a property, while retaining the folding of the polypeptide backbone.

scanning force microscopy: (= atomic force microscopy (AFM))

scanning: In protein synthesis, the movement of a ribosome along an RNA molecule. Also, the rapid surveying of clones, libraries or DNA sequences for a desired trait. (*see also* genomic mismatch scanning (GMS); homologue-scanning mutagenesis; PCR-based differential scanning)

Scatchard plot: A graphical method for determination of a binding constant and the number of binding sites. The ratio of the amount of ligand (e.g. a hormone) that is bound (e.g. to a receptor) to the amount that is free is plotted against the free amount.

scavenger: A compound that can combine with a free radical and thus alter the kinetics or course of free-radical reactions. (*see also* antioxidant)

Schardinger dextrin: A product of bacterial degradation of starch; a closed-ring oligosaccharide in which glucosyl monomers are joined by a14 glycosidic bonds.

Schiff base: The product formed by the reversible condensation of an aldehyde with an amino group in which the elements of water have been eliminated. It is featured in structures of biochemical interest, notably the reactions of amino acids with pyridoxal phosphate as a coenzyme of various enzymes of amino acid metabolism.

Schiff's reagent: (*see* periodic acid-Schiff (PAS) stain)

schlieren method: An optical method for detection of the non-uniform dispersion of a macromolecule in solution, especially during analytical ultracentrifugation or free-zone electrophoresis, by the refractive index of its solution.

scintillation counting: A technique for detection and quantification of radioactivity by capture of the emission from a radiolabel by a primary phosphor which, on emission, excites a secondary phosphor and causes it to emit a pulse of detectable visible light. (*see also* excimer)

scintillation proximity assay (SPA): A method for quantification of a molecule on the basis of its containing a specific ligand. The molecule is labelled with a weakly emitting radioisotope and a ligand, e.g. a [3H]glycoprotein. The molecule becomes adsorbed to minute beads that contain a scintillant that is excited by the energy of radioactive decay and a recognition site for the ligand, e.g. concanavalin A. Secondary emission of light from the scintillant is quantified in a scintillation counter. Because a weakly emitting radioisotope is used, only when the radioactive decay occurs in close proximity to the scintillant (i.e. when the molecule is bound to the bead) is a visible emission produced.

scissors grip: A term borrowed from wrestling terminology to describe the interaction of a bZIP protein with the double-standed DNA to which it binds.

sclero-: A prefix derived from the Greek for hard.

scRNA: Small cytoplasmic RNA.

SDS gel electrophoresis: (= sodium dodecyl sulphate/polyacrylamide-gel electrophoresis (SDS/PAGE))

SDS/PAGE: (= sodium dodecyl sulphate/polyacrylamide-gel electrophoresis (SDS/PAGE))

seco-: A prefix that signifies a product of the opening of a ring, e.g. a product of A-ring opening, a 2,3-seco-steroid, or if the ring-opening is oxidative, a 2,3-seco-dioic acid.

second genetic code: An imprecise term that sometimes refers to the nature of the amino acid residues of a protein which determine its secondary and tertiary structure, and sometimes to the features of a tRNA molecule that make it recognizable by one amino acid synthetase but not by others.

second law of thermodynamics: The relationship of free energy (G), enthalpy (H) and entropy (S) when a system changes from one equilibrium state to another: G=H-TS, where T is the absolute temperature. The law provides the conceptual basis for redox and group transfer potentials; for example, the Nernst equation or the equilibrium constant, which is G=-RTln[Keq], where R is the gas constant. The law allows prediction of the direction in which reversible reactions may proceed, dependent upon reactant and product concentrations, by quantifying and defining free energy for a spontaneous chemical change as a negative value which is independent of the path between the initial and final state. Using transition state theory, the law may be applied to rate constants and their temperature dependencies.

secondary metabolism: Those metabolic pathways, other than those for energy production and for biosynthesis of nucleic acids, proteins, structural components, etc., that are unique to an organism; especially those pathways by which bacteria, moulds and plants synthesize pigments, antibiotics, toxins and other natural products.

secondary structure: In protein chemistry, the regular folding of a polypeptide in a repeated pattern, e.g. -helix, -pleated sheet, -turn; in nucleic acid chemistry by analogy, the double-helical structure of a polynucleotide and other regular structures seen in RNA foldings. (*see also* primary structure; quaternary structure; supersecondary structure; tertiary structure; foldamer)

secretagogue: A chemical agent that promotes secretion.

secretory vesicle: A membrane-bound organelle near the apical membrane of a cell that contains a protein, neurotransmitter or other substance awaiting a signal for secretion. (*see also* zymogen granule)

sedimentation constant: A measure of the rate of sedimentation in an analytical ultracentrifuge; v/2x, where v is the rate of sedimentation, is the angular velocity in radians/s and x is the distance from the axis of rotation; usually expressed in Svedberg units (10-13s).

sedimentation-velocity ultracentrifugation: A technique that measures the rate of movement of a dissolved polymer under the force exerted upon it in a centrifuge, to assist in evaluation of molecular mass, axial ratio and other hydrodynamic properties. (*see also* equilibrium sedimentation ultracentrifugation)

segmental flexibility: The internal movement possible in some proteins, in which domains show little internal movement but move relative to each other by flexion at the hinges that connect them, e.g. the wagging movement of the antigen-binding sites of immunoglobulin G towards or away from each other.

selected and amplified (protein) binding site oligonucleotide: (*see* cyclic amplification and selection of targets (CASTing))

selective pressure: In evolutionary theory, the effect on survival of a species of the sum of all factors, physical and behavioral, inherent and environmental; especially as an inherited trait may marginally effect survival under the influence of these factors.

selective theory (clonal selection theory): A hypothesis that explains antibody specificity by the existence of a wide variety of antibodies, even before exposure to any particular antigen. (*see also* instructive theory)

SELEX: Systematic evolution of ligands by exponential enrichment. (*see* cyclic amplification and selection of targets (CASTing))

self amplification: *see* bidirectional PCR amplification of specific alleles (bi-PASA)

self-absorption: *see* atomic absorption spectrometry

self-assembly: The spontaneous aggregation of a complex from its components, the chemistry of which determines the nature of the complex, e.g. the formation of a lipid bilayer from phospholipid molecules.

selfish gene: A gene which confers a selective advantage upon its own propagation, but not on the survival of the individual. A classic example is Segregation Distorter in D. melanogaster: in heterozygous flies, sperm containing the chromosome with the gene survive but sperm with the chromosome without the gene are killed (homozygous flies with the gene are sterile).

self-splicing: The activity of the precursor of mature RNA whereby it catalyses its own (cis>) splicing. Group I introns (examples are Tetrahymena rRNA and hammerhead ribozymes) are self-spliced with the assistance of Mg2+ ions, an internal guide base sequence near the 5'-end and other sequences that form internal base pairing to bring the structure into a reactive supersecondary structure; reaction proceeds by attack of the 3'-hydroxyl group of a guanosine (or guanine nucleotide) on the upstream exon-intron junction, which displaces the 3' end of the upstream exon; its newly generated 3' end then attacks the junction between the intron and the downstream exon to generate the spliced exons and the free intron. Group II intron self-splicing differs in that the attacking agent is the 2'-hydroxy group of an adenylate residue of the intron located near its junction with the downstream exon, and intermediates are the upstream exon and the downstream exon still ligated to the intron, whose 5'-guanosine hydroxy group is looped back to the internal adenylate residue and is held in a 2',5' phosphodiester bond. The looped intron (a lariat) is finally displaced by attack of the 3'-hydroxy group of the upstream exon on the intron-(downstream) exon junction. Spliceosomes use the group II mechanism, but as they include ancillary RNA and protein molecules, they are not in a strict sense, self-splicing. The phenomenon of trans-splicing refers to ligation of the upstream site to a downstream site which is not the closest in the linear sequence.

self-sustained sequence replication (3SR): A technique for amplification of a DNA sequence, based on the steps of retroviral replication of nucleic acids. Beginning with a clean DNA sequence, DNA-dependent RNA polymerase makes a RNA transcript; from this, a DNA transcript is made by reverse transcriptase. Because the RNA of the RNA-DNA hybrid which is produced by the reverse transcriptase is degraded by RNaseH, amplification is linear, not exponential. (*see* polymerase chain reaction; *see* strand displacement amplification)

semi-conservative replication: The mode of DNA synthesis that results in each new duplex having one parent polynucleotide strand and one newly synthesized strand. (*see also* Meselson-Stahl experiment)

semi-discontinuous: Descriptive of the replication of DNA in which the parent strand that is exposed from its 3'-end towards its 5'-end is used as a template for continuous synthesis of a new strand, but the parent strand that is exposed from its 5'-end towards its 3'-end is used as a template for synthesis of short discontinuous segments of DNA, called Okazaki fragments.

semiochemical: Referring to chemicals, especially those in the environment, which effect physiological changes; e.g. pheromones, odorants.

semi-permeable: Descriptive of a membrane that will permit transport of solvent and small solute molecules but will retain larger solutes. (*see also* dialysis)

semiquinone: A free-radical intermediate between a quinol and the more oxidized quinone.

sense strand: The strand of double-stranded DNA that contains genetic information; the template or coding strand; often represented as the (+)-strand. (*see also* antisense)

sensitivity: In quantitative terms, the responsiveness of a physiological system to a stimulus, such as an allosteric effector, substrate or hormone. Where X is the magnitude of the stimulus and Y the magnitude of the response, sensitivity can be expressed as the dimensionless relative increase of Y per relative increase in X, effectively the slope of a plot of logY against logX. Alternatively, sensitivity is expressed as the increase in Y per increase in X, i.e. dY/dX, which has the dimensions of the dependent and independent variables.

sequenator: A device to sequentially degrade a protein to determine its primary structure.

sequence insertion: (*see* gene targeting)

sequence replacement: (*see* gene targeting)

sequence tagged site: (*see also* comparative anchor tagged site; expressed sequence tag)

sequence: The ordered linear array of amino acid residues in a protein, of bases in a polynucleotide, or of glycosides in a complex polysaccharide; also, as a verb, to determine such a sequence, e.g. to sequence a protein.

sequencing: The determination of the base sequence of a homogeneous DNA fragment, the amino acid sequence of a protein, or the glycoside sequence of a complex polysaccharide.

sequencing by hybridization: A proposed approach to rapid sequencing of genomic DNA. The key tool for analysis is a superchip, a surface over which are attached at minute yet identifiable positions unlabelled oligonucleotide probes. For example, if pentanucleotides are used, there will be 625 (54) rows of unique pentanucleotides. The entire surface of the superchip is exposed to an amplified sample of the test DNA plus a ligase. Each unlabelled probe will hybridize to the DNA where there is a complementary sequence. The superchip is then overlaid by another chip that contains a similar array of rows of labelled oligonucleotides, but with the rows running across the rows of unlabelled probes. Where the labelled and unlabelled probes are complementary to adjacent sequences (in the example, a decanucleotide sequence), the ligase can join them and thus fix the label to the superchip. A robotic reader will identify the positions on the superchip that have become labelled and that therefore indicate the presence of a particular (decanucleotide) sequence. The overlap of the identified sequences will allow reconstruction of the entire sequence of the test DNA.

sequential: (*see* enzyme mechanism)

sequential feedback: A control mechanism for a branched metabolic pathway. The end product of each branch inhibits the first enzymic step of its own branch, which results in an accumulation of the last metabolite of the common pathway. This causes inhibition of the first step of the common pathway and results in inhibition of the common pathway if both end products are present in excess.

sequential model: (= Koshland model (KNF model))

serine proteinase: A type of peptidase that has at its active site a serine residue. (*see also* aspartate proteinase; cysteine proteinase (thiol proteinase); metalloproteinase)

sero-: A prefix derived from the Latin for whey. (*see* serum)

serotype: An immunologically defined variant of a micro-organism; one that bears a unique epitope on its surface.

serpentine receptor: A large family of integral membrane proteins with seven transmembrane helical domains; on the extracellular surface they bind activating ligands, which induce conformational changes through to the intracellular surface, where they interact with G protein subunits. *see* seven transmembrane domain proteinserpin. Serine proteinase inhibitor; one of a group of homologous proteins, e.g. 1-antiproteinase.

serum: The liquid phase that remains after blood has clotted. (*see also* plasma)

sesquiterpene: (*see* terpene)

seven transmembrane domain protein (7TM): One of a group of structurally related long polypeptide chains which possess seven a-helical domains that span a membrane, the intervening intra- and extracellular loops which connect them, and the polypeptide's intracellular C- and N-terminal sequences. *see* serpentine receptor *see* transmembrane domain

seven-helix motif: A characteristic of several receptor proteins, that zig-zag seven times as -helices across the plasma membrane. (*see also* seven transmembrane domain protein (7TM))

sex pilus: In bacterial conjugation, an appendage of a male bacterium by which it attaches to a female bacterium, preparatory to the transfer of DNA from male to female.

shadow band: *see* slippage

shape library: An assortment of oligonucleotides, often of random sequence, some of which (or some of the encoded proteins of which) can bind to secondary structures of other oligonucleotides or proteins. The common features of the selected polymers are termed the recognition consensus sequence (RCS).

sheddase: = membrane protein secretase

shikimic acid metabolite (C6C3 metabolite): A derivative of shikimic acid with an aromatic 6-carbon ring and an attached 3-carbon side chain, e.g. cinnamic acid, phenylalanine.

Shine-Delgarno sequence: A polynucleotide sequence of prokaryotic mRNA that is complementary to, and therefore base-pairs with, a sequence at the 3'-end of bacterial 16S rRNA, which function to position the initiator codon at the peptidyl site of the ribosome.

short tandem repeat (STR): A 2-6bp sequence in genomic DNA that is repeated a number of times that is characteristic of an individual. (*see also* variable number tandem repeat (VNTR))

short-chain fatty acid: (*see* medium-chain fatty acid)

short-column method: An analytical ultracentrifugation technique for measurement of the molecular mass of a protein; requires relatively brief periods of time because it uses very short columns of liquids since the distance from meniscus to base is 3mm or less.

shotgun sequencing: An approach to determination of the sequence of a large, randomly selected, piece of genomic DNA. A cosmid that contains the DNA is sonicated into smaller fragments of 1.0-1.5kb, which are ligated into a vector (e.g. M13) containing a universal primer, and each fragment is sequenced. In order to obtain the complete linear structure of a 40kb cosmid, 800-1000 fragments must be sequenced. More recently, bacterial artificial chromosomes (BACs) have been employed, allowing up to 350kb fragments to be cloned for sequence analysis.

shrinkage temperature: The temperature at which collagen fibres contract; analogous to the melting temperature of a protein.

shuffling: Recruitment of the exons that code for a functional domain of one protein in the evolution of a gene for a new protein, to produce hybrids with recognizable motifs, such as kringles and metal-binding fingers. A consequence of this hypothesis is the corollary that introns are as old as structural genes; this is upheld by the intron-early school but opposed by the intron-late school, who point to the absence of introns in prokaryotes and some primitive eukaryotes.

shunt: Usually, the pentose phosphate pathway.

shuttle: A mechanism for transport of metabolites or chemical groups across the mitochondrial membrane, e.g. the electron shuttle that transports glycerophosphate into, and dihydroxyacetone phosphate out of, the mitochondrion, which permits oxidation of NADH in the cytosol and reduction of FAD inside the mitochondrion.

shuttle vector: A plasmid that has both bacterial and eukaryotic origins of replication and so can propagate in either kind of cell; useful as a form of recombinant DNA for growth in a bacterium and subsequent transfer into eukaryotic cells for expression.

side chain: The moiety of an amino acid residue in a protein, or of a free amino acid, that is attached to the -carbon and is unique to each amino acid, e.g. the isopropyl group of valine, the benzyl group of phenylalanine. The group attached to C-17 of sterols, mineralocorticoids and bile acids is also called a side chain.

side shelf experiment: An experimental approach to discovery of a metabolic precursor or regulator, in which candidates for activity are randomly selected according to availability rather than by intelligent design.

sidero-: A prefix derived from the Greek for iron.

siderophore: A low-molecular-mass compound that binds to ferric ions and facilitates their absorption by the micro-organisms that produce them; e.g. enterobactin, the catechol derivative of some bacteria, and ferrichrome, the hydroxamate compound of some fungi.

sigma (Σ) subunit: The subunit of RNA polymerase that recognizes transcription initiation sites on double-stranded DNA.

sigmoidal: S-shaped; descriptive of the appearance of a graph that shows dependence of rate on substrate concentration for a multimeric enzyme that is accelerated by binding of more than one substrate molecule; also descriptive of non-simple binding of ligands to a non-enzymic protein, e.g. oxygen binding to haemoglobin.

signal hypothesis: The proposal that secretory proteins are synthesized with N-terminal signal sequences, which assist their passage into secretory vesicles. (*see* leader sequence)

signal joint: *see* V(D)J recombination

signal molecule: A neurotransmitter, hormone, second messenger or other regulatory molecule.

signal patch: An area of a protein surface that is recognized by the sorting mechanism of a cell and allows the protein to be directed to its appropriate locus within the cell.

signal recognition particle (SRP): In the synthesis of secretory, endoplasmic reticulum (ER), Golgi apparatus, lysosomal and integral membrane proteins, a ribonucleoprotein that recognizes a signal sequence and prevents further translation until the docking protein (SRP receptor) of the ER can direct the growing polypeptide across the membrane of the ER. (*see also* ribophorin)

signal sequence (leader sequence): The N-terminal portion of a secretory or membrane protein that assists it across the membrane of the rough endoplasmic reticulum, where it is synthesized, but is cleaved from the protein even before the synthesis of the protein is complete. Mitochondrial and chloroplast proteins that are synthesized on ribosomes also have signal sequences that target them to their organelle; these may be N-terminal or internal, and may or may not be removed by proteolysis.

signal transducer and activator of transcription: *see* STAT

signal transduction: *see* preassembled signalling complex

signature sequence: An insertion or deletion in the coding sequence of a gene which, because it is shared by phylogenetically distant species, is thought to be evolutionarily conserved and therefore can serve to trace evolutionary relationships of species.

signature sequence tag (SST): one of a set of short DNA sequences for which probes are available for the identification of a gene. As the combination of 600 to 1500 SSTs provide a unique signature for any gene, such probes can be used as an alternative to expressed sequence tags as a means of detection of a gene in cDNA libraries.

silencer: (*see* silencing)

silencing: The genetically programmed repression of transcription. Silencing may be the net effect of regulatory events that increase transcription and those which repress it (the analog model), or it may be an all-or-none phenomenon in which the gene exists with its promoter and enhancer sites open and its coding region available to transcription or tightly folded and packaged into heterochromatin (the digital or binary switch model). In the latter view, the switch is operated by two similarly acting sequences: a locus control region (LCR) and a silencer. Both are cis-acting, function bi-directionally and their actions are position and distance dependent. The LCR promotes unfolding of the chromatin and allows a promoter to act; the silencer has the opposite effect. The switch is apparently set in each cell cycle during the S-phase. (*see also* imprinting; origin recognition complex)

silent mutation: A mutation that causes no functional change in the gene product. (*see also* transparent mutation)

simple sequence DNA: The highly repetitive non-coding DNA of a satellite that serves a structural role in mitosis; used for genetic characterization by its polymorphism (simple sequence length polymorphism, SSLP).

simplified: Descriptive of a protein whose amino acid sequence has been modified to minimize its complexity, e.g. by substitution in many sequence positions with only a few selected amino acids.

SIMS: Secondary-ion bombardment mass spectrometry. (*see* mass spectrometry (MS))

single cell gel electrophoresis (SCGE): = comet assay

single locus DNA profiling: The use, e.g. in forensic science, of a single variable number tandem repeat to characterize an individual's DNA.

single-domain antibody (dAB): The variable domain of an immunoglobulin heavy (VH) or light (VL) chain, each of which can have considerable affinity for an antigen; produced in bacteria by biotechnological methods, whereby the newly synthesized protein can pass freely through the cell wall into the medium.

single-lane sequencing: (*see* fluorescence-based DNA analysis technology (FDAT))

single-strand conformational polymorphism (SSCP): A technique for detection of mutations in a defined DNA sequence. A labelled PCR-amplified segment of DNA is heat-denatured then quenched at 08C, and the resultant single-stranded polynucleotides are electrophoretically separated on non-denaturing gels. Intrachain base pairing results in a limited number of conformers stabilized by intrachain loops, and mutated DNA shows on electrophoresis an altered assortment of such conformers. (*see also* restriction endonculease fingerprinting (REF))

singlet state: An excited electronic state in which an electron is raised to a higher energy level without reversal of its spin state, and thus is easily able to fall back to its ground state, often with a fluorescent emission. (*see also* triplet state)

singlet strand: An unpaired polynucleotide chain.

single-well hydrogen bond: (*see* hydrogen bond)

sirohaeme: A prosthetic group of some redox proteins, consisting of a reduced porphyrin and a Fe4S4 iron-sulphur centre.

site-directed mutagenesis: A technique to alter deliberately a structural gene to insert a desired point mutation. A synthetic polynucleotide that contains the genetic alteration is used as an imperfectly fitting primer in replication and the completed (-)-strand incorporates the mutation.

site-specific mutagenesis: (= site-directed mutagenesis)

slab gel: A thin sheet that acts as a support for electrophoresis, often composed of polyacrylamide or agarose. (*see also* disc-gel electrophoresis)

sliding filament model: A conceptualization of muscle contraction in which thick filaments in the centre of a sarcomere are surrounded by thin filaments that are attached to the ends of the sarcomere and extend towards its centre. Stimulation drives the overlap of the thick and thin filaments and thus pulls the ends of the sarcomere towards the centre.

slippage: A source of microsatellite polymorphism. During DNA replication, the polymerase periodically dissociates from the site of synthesis of the lagging strand. If this occurs in a microsatellite sequence, a region of dinucleotide repeats, the 5'-end of the lagging strand may momentarily dissociate from the template and incorrectly reassociate with a dinucleotide loop in the newly synthesized strand or in the template. When the polymerase complex reassociates and continues synthesis, an extra dinucleotide may be inserted if the loop had been in the lagging strand, or deleted, if the loop had been in the template. When amplified and electrophoresed, sequences with such polymorphism display shadow or stutter bands that differ by 2 bp.

slow-twitch muscle: (= red muscle (slow-twitch muscle))

small G protein: A class of proteins that effect intracellular structural changes via modifications of the cytoskeleton. Like other G proteins, they are active when bound to GTP, but inactive when the GTP is exchanged for a GDP. Activity is regulated by proteins which affect the GTP-GDP exchange.

small interspersed repeat element (SINE): A group of recurring species-specific polynucleotide sequences in the human genome, e.g. an Alu sequence of 0.3kb present in approximately 106 copies. (*see also* long interspersed repeat element (LINE))

small molecule library: A collection of related structures synthesized for screening for a desired property. The library may by synthesized by parallel, split or spatially addressable synthesis. Parallel synthesis is a single-batch method that uses a mixture of reagents at each step of the synthesis to generate the diversity of products. Split synthesis requires a solid-phase support to allow separation of products at each synthetic step. For example, a polymer may be assembled by coupling monomers A, B and C to beads; the separate batches are filtered and combined, then divided again into thirds for coupling with monomers D, E and F. After n cycles, there will be 3n sequences present. Spatially addressable synthesis localizes the reactions on specific locations on a chip, a solid surface ruled into a grid, each locus on which receives a unique sequence of reagents. The products of each synthetic method are exposed to a selective procedure. The position on a chip directly identitfies the structure assembled there. If a binding property is sought, localization to a ligand on a solid-phase surface will assist selection from a library of dissolved structures. Coding of beads during their synthesis allows their identification after selection; such coding may be the parallel assembly of a protein or nucleic acid during each step of the synthesis, polymers of which are easily sequenced by available technologies.

small nuclear RNA (snRNA): A category of polynucleotides, usually from 70 to 500 nucleotides in length. As complexes with proteins (snrps) they have several functions, among them the processing of pre-mRNA, either directly as part of a spliceosome or indirectly by hybridizing with the pre-mRNA and altering its architecture, i.e. by altering its secondary structure, its tertiary and quaternary structure is also altered. Some snrps are encoded by their own genes, others by the introns of a host gene which itself has some ribosome-related function. (*see also* spliceosome)

small nucleolar RNA (snoRNA): One of several oligonucleotides of the nucleolus that have roles in pre-mRNA processing.

smart screening: The use of bioinformatics for discovery of useful correlations; e.g. for the determination of genes that are uniquely expressed by osteoclasts, in order select a target for the design of an osteoporosis drug.

smooth endoplasmic reticulum: (= endoplasmic reticulum (ER); *see* also rough endoplasmic reticulum (RER))

SMP: (= submitochondrial particle (SMP))

snap-back sequence: Half the base-pairing sequence of a hairpin turn of a tRNA, which can easily rebind to its complementary sequence after denaturation.

SNARE hypothesis: A proposal for the mechanism by which membranes, particularly vesicular and Golgi or plasma membranes, fuse during, for instance intracellular transport and secretion. The two membranes contain protein complexes, SNAREs, which will become the sites of fusion. The regretable nomenclature derives from an N-ethylmaleimide-sensitive (i.e. sulphydryl-containing) soluble fusion protein (NSF) and soluble NSF attachment proteins (SNAPs); the membrane-associated complexes that recognize SNAPS are soluble NSF attachment protein receptors (SNAREs). The vesicular membrane carries a v-SNARE and the target membrane, a t-SNARE. The two SNAREs can form a fusion particle, and with energy input from the hydrolysis of ATP, can cause membrane fusion. Spontaneous fusion is prevented by v- and t-SNARE clamping proteins.

sniffer-patch: A biosensor for the detection of a ion channel- activating neurotransmitter. A small area of cell membrane containing receptors for the neurotransmitter closes the end of a micropipet which acts as an electrode. This is then put into proximity of a cell suspected of releasing the neurotransmitter. In the presence of the neurotransmitter, the membrane patch will activate the ion channels, and the flux of ions will be detected by the electrode.

snRNA: (= small nuclear RNA (snRNA))

snrp: Small nuclear RNA-protein complex. (*see* small nuclear RNA (snRNA); *see also* pre-mRNA splicing)

snuRNA: Small nucleolar RNA.

soap: A sodium, potassium or other soluble salt of a fatty acid. sodium dodecyl sulphate/polyacrylamide-gel electrophoresis (SDS/PAGE). A technique to dissociate multimeric proteins and to separate all proteins according to their apparent molecular masses. Proteins are treated with sodium dodecyl sulphate (SDS), an anionic detergent, and separated by electrophoresis on polyacrylamide gels that incorporate SDS. The detergent adsorbs on to the proteins in proportion to their mass and dominates their electric charges. Also known as denaturing gel electrophoresis or SDS gel electrophoresis. (*see also* native gel).

sodium dodecyl sulphate/polyacrylamide-gel electrophoresis (SDS/PAGE): A technique to dissociate multimeric proteins and to separate all proteins according to their apparent molecular masses. Proteins are treated with sodium dodecyl sulphate (SDS), an anionic detergent, and separated by electrophoresis on polyacrylamide gels that incorporate SDS. The detergent adsorbs on to the proteins in proportion to their mass and dominates their electric charges. Also known as denaturing gel electrophoresis or SDS gel electrophoresis. (*see also* native gel)

solenoidal model: A representation of the organization of chromatin in which adjacent nucleosomes in the linear DNA sequence are wound into a 36nm diameter helix that contains six nucleosomes per turn.

solenoidal: Descriptive of one form of slightly unwound supercoiled DNA in which the double helix lies in coils that are stacked one upon another in a very condensed arrangement. (*see also* solenoidal model)

solid-phase sequencing: A set of methods for sequencing of proteins and synthesis of proteins and nucleic acids, in which the subject molecule is attached to a solid support to aid in its separation from reagents during chemical reactions. The reagents may be in a liquid or a gas phase. In one procedure, the reaction takes place at the solid-liquid interface of a spinning cup that holds the immobilized polymer and the reagent solution.

soluble NSF attachment protein (SNAP): (*see* SNARE hypothesis)

soluble RNA: (= tRNA)

solvation substitution: (*see* mobile barrier)

solvent-accessible surface: A feature of a macromolecule defined by a molecular modelling technique, e.g. that of Connolly (a Connolly surface); the parts of the structure that are exposed to the solvent molecules are identified by rolling a sphere the size of a water molecule over its surface.

solvent isotope effect: The effect on an enzyme's kinetics of inclusion of 2H2O; often expressed as Dk, the ratio of the reaction rate constant in H2O to the rate constant in 2H2O. An extension of this approach is a proton inventory, in which the rate constant of each phase of an enzymic transformation is determined as a function of the mole fraction of 2H2O, in order to determine the number of protons transferred in each discrete step of the catalytic cycle.

soma-: A prefix derived from the Greek for body, which indicates an anatomically distinct structure.

somatic cell: A non-gamete.

somatic cell hybrid mapping: (*see* mapping; radiation-induced hybrid mapping)

somatic gene therapy: (*see* gene therapy)

somatic hybrid: The fusion product of two different types of non-gamete cells.

sonication: A technique to disrupt cell structures by subjecting them to high-frequency sound waves.

Soret band: A spectrophotometric characteristic of porphyrins; a strong absorption band near 400nm.

sorting: The translocation to their appropriate loci within a cell of newly synthesized or endocytosed proteins and nucleic acids.

SOS repair: An error-prone DNA repair process induced in response to ultraviolet-light-induced mutagenesis. In the face of pyrimidine dimerization, it allows for continued DNA resynthesis with relaxation of the strict requirement for base pairing.

Southern blotting: A technique for detection of specific DNA fragments. Restriction fragments are separated by polyacrylamide-gel electrophoresis, blotted on to a plastic sheet (e.g. of nitrocellulose) and detected by hybridization with a radiolabelled probe followed by autoradiography. (*see also* electroblotting; northern blotting; south-western blotting; western blotting (immunoblotting))

south-western blotting: A technique for identification of DNA-binding proteins and regions of DNA that bind proteins which combines aspects of Southern and western blotting techniques. A nuclear extract is separated by sodium dodecyl sulphate/ polyacrylamide-gel electrophoresis, renatured and blotted on to nitrocellulose, then exposed to labelled DNA probes. Photographic film laid against the gel is exposed so as to identify which protein bands have bound DNA probes, and then the DNA is extracted from them. (*see also* Southern blotting; western blotting (immunoblotting))

SPA: (= scintillation proximity assay (SPA))

spacer arm: In affinity chromatography, the chain of carbon and/or other atoms that positions a functional group away from the solid matrix to which it is covalently bound and makes it more available to a ligand and less restricted by steric hindrance by the matrix.

spacer DNA: DNA that does not yet have a recognized function. (*see also* junk DNA)

spare receptors: The phenomenon of a hormone-responsive cell having more receptors capable of binding the hormone than are needed to effect a maximal response; possibly an adaptation that makes the cell responsive to the rate at which the hormone can reach the cell surface (the collisional limit).

spatially addressable synthesis: (*see* small molecule library)

SPCR: (= synthetic peptide combinatorial library (SPCL))

species: A group of organisms that breed among themselves to produce viable offspring.

specific acid: In chemistry, a hydronium ion that can participate in catalysis. (*see also* general acid; specific base)

specific activity: A measure of function per unit, e.g. enzyme activity per mg of protein, or radioactivity per weight (or per volume, or per mole) (more correctly known as specific radioactivity).

specific base: In chemistry, a hydroxyl ion that can participate in catalysis. (*see also* general base; specific acid)

specific dynamic action: (= thermic effect of food (TEF))

specific linking difference: (= superhelical density (specific linking difference))

specific radioactivity: A measure of the degree of labelling with a radioactive tracer; sometimes referred to as specific activity; expressed as radioactivity per unit mass, e.g. Ci/mmol, c.p.m./mg.

specificity constant (kcat/Km): The apparent second-order rate constant for enzyme action under conditions where an enzyme's binding site is largely unoccupied by substrate ([S]Km); equivalent to the slope of the (pseudo) first-order region of a plot of enzymic rate against substrate concentration.

specificity subsite: A part of the active site of a peptidase. Amino acid residues adjacent to the scissile bond are designated P1, P2, P3, etc. on the N-terminal side, and P1', P2', P3', etc. on the C-terminal side; P1-P1' is the scissile bond. Peptidase subsites that bind these are designated S1, S1', S2, S2', etc.

spectral karyotyping: The colour-coded display of metaphase chromosomes. Probes which are unique to each chromosome are constructed by separation of individual chromosomes and their fragmentation into 400 to 500 pieces and labelling with a combination of fluorophores that is unique to each chromosome. Metaphase chromosomes are hybridized with a mixture of these probes and their microscopic image is captured by a charged coupled device (CCD) camera in several tens of thousands of pixels. The fluorescence emission spectrum of each pixel is analysed to identify the combination of fluorophores from which it is generated and thereby identify the chromosome to which it belongs; each pixel then can be assigned to a chromosome. Each chromosome, therefore each pixel, is arbitrarily assigned its own characteristic colour in order to display the colour-coded chromosome. Translocations and insertions of chromosome fragments are seen as chromosomes with bands of more than one colour. *see* coincidence painting (chromosome painting)

spermatozoon: A male gamete, characterized in part by an acrosome and a flagellum.

spheroplast: A plant cell that has been incompletely stripped of its cell wall. (*see also* protoplast)

spin imaging: (= zeugmatography (spin imaging))

spin label: A chemical functional group with an unpaired electron attached to a compound or macromolecule, e.g. a nitroxide group; detectable and characterizable by its electron spin resonance spectrum.

spin lattice: The magnetic environment that surrounds the nucleus or electron under observation in nuclear magnetic resonance or electron spin resonance studies.

spindle: The microtubule structure that radiates from each pole of a cell to the chromatids during mitosis or meiosis.

spliceosome: A complex of (small nuclear RNA)-protein complexes and other proteins that assemble on a pre-mRNA and catalyse the excision of an intron in a process mechanistically similar to group II self-splicing. It acts presumably by forcing the intron into a loop and bridging the pre-mRNA at its splicing sites.

splicing: he process by which an intron is excised and exons are religated in the post-transcriptional modification of RNA (cis-splicing); also the excision of an intein from a precursor protein.(*see also* protein splicing; self-splicing; spliceosome)

splinkerette PCR: (*see* vectorette PCR)

split gene: A gene whose coding sequence is interrupted by introns.

split synthesis: (*see* small molecule library)

splitting: The generation of multiple signals in nuclear magnetic resonance and electron spin resonance spectroscopy due to the interaction of nuclei or electrons coupled through space. (*see also* hyperfine splitting; Zeeman splitting)

SPR: Surface plasmon resonance. (*see* biomolecular interaction analysis (BIA))

spreading factor: An agent that permits the diffusion of foreign materials through tissue, e.g. hyaluronidase, which degrades the extracellular matrix.

squelching: (*see* trans-acting)

src gene: A polynucleotide that in one form, c-src, is a normal cellular gene, and in a similar form, v-src, is an oncogene of avian sarcoma virus.

src protein: A product of the src gene; acts to phosphorylate tyrosine residues of some proteins.

sRNA: Soluble RNA. (= tRNA)

SRP: (= signal recognition particle (SRP))

SRP receptor: (= docking protein; *see* signal recognition particle (SRP))

SSCP: (= single-strand conformational polymorphism (SSCP))

ssDNA: Single-stranded DNA.

ssRNA: Single-stranded RNA.

staggered ended: (= sticky ended)

standard state: For solutions in general, the conditions under which thermodynamic properties of a solute are defined, namely 1M. The properties of the solution are those extrapolated from very dilute solutions. In biochemistry the standard state is defined for [H+]=10-7M (pH 7), and thermodynamic properties of all other ionizing species are defined for a total of all protonated and deprotonated species of 1M.

star activity: A secondary nucleotide sequence specificity of an endonuclease. After the canonical specificity is established, a library of potential substrates may be created by substituting at each sequence position, all of the other three nucleotide residues. Each such set is written with an asterisk (star) in the place of the substituted nucleotide residue.

start signal: In translation, the initiation codon (AUG); in transcription, the site of initiation of RNA synthesis.

state 1, state 2, etc.: (*see* respiration state)

STC: (= sequence-tagged connector (STC))

steady state: In kinetics, the maintenance of the concentration of an intermediate by its formation from precursors at the same rate as its conversion into products.

stefin: (*see* cystatin)

stem cell (blast cell): A relatively undifferentiated cell that can develop into a more specialized cell.

stereo-: A prefix derived from the Greek for solid, which refers to relationships in three-dimensions.

stereochemical code: The dependence of the secondary and tertiary structure on the primary structure of a protein, which allows the prediction of three-dimensional structure from sequence information.

stereoelectronic control: (*see* entropy effect)

stereoisomer: One of two or more compounds that differ only in their orientation at one or more asymmetrical centres. (*see also* diastereoisomer; enantiomer; epimer)

stereoisomer: One of two or more compounds that differ only in their orientation at one or more asymmetrical centres. (*see also* diastereoisomer; enantiomer; epimer)

stereopopulation control: (*see* entropy effect)

stereospecific transfer: The stereospecificity of an enzyme that catalyses either the abstraction of an atom or group from a prochiral centre, or the addition of an atom or group to form one.

steric hindrance: The constraint on a reaction or conformational change that is due to crowding of atoms within the van der Waals radii of other atoms.

steroid receptor: (*see* nuclear receptor)

steroid: A derivative of perhydrocyclopentanophenanthrene; a more oxygenated product of cholesterol, with the C-17 side chain shortened or removed. Many steroids are hormones, e.g. cortisone, progesterone, oestradiol. (*see also* sterol)

sterol: A crystalline alcohol that incorporates a characteristic perhydrocyclopentanophenanthrene ring system and a branched hydrocarbon side chain, e.g. cholesterol, ergosterol. (*see also* steroid)

sticky ended: Descriptive of the structure of a double-stranded polynucleotide in which one strand extends further from the end than the other to create a single-stranded 'tail'; especially the product of cleavage by a restriction endonuclease at a palindromic sequence. Also known as cohesive ended or staggered ended. (*see also* blunt ended (flush ended))

stochastic: Determined by the laws of chance and probability.

Stokes radius: On the assumption that it is a sphere, the apparent radius of a macromolecule in solution, as determined from its hydrodynamic behaviour, i.e. intrinsic viscosity, diffusion or sedimentation coefficient, or behaviour in gel filtration chromatography. (*see also* Stokes' Law)

Stokes' Law: A theoretically derived relationship between the viscosity coefficient, , the frictional coefficient, f, and the radius of a macromolecule (the Stokes radius) r: f=6r. This calculated f may be compared with that determined experimentally from the relationship between the diffusion coefficient, D, the Boltzmann constant, k, and the absolute temperature, T: D=kT/f. The deviation (usually an increase) of the measured f from the calculated f is a measure of the solvation of the macromolecule or its deviation from a spherical shape.

stopped-flow: A rapid mixing technique for observation of fast reactions, usually in the millisecond range. Two syringes holding solutions that must be mixed to initiate the reaction are emptied by a single piston that forces the contents through a mixing chamber into an observation chamber where spectral properties are observed and stored in an oscilloscope for subsequent analysis.

stop-transfer sequence: A polypeptide sequence of a nascent membrane protein that prevents its translocation into a transfer vesicle and permits its insertion into a membrane.

store-operated calcium entry (SOCE): The phenomenon of replacement of the endoplasmic reticulum Ca2+ stores from the extracellular Ca2+ pool following mobilization of Ca2+ by cell stimulation.

store-operated calcium entry: The phenomenon of replacement of the endoplasmic reticulum Ca2+ stores from the extracellular Ca2+ pool following mobilization of Ca2+ by cell stimulation.

strand displacement amplification (SDA): A technique for amplification of DNA sequences. The source DNA is denatured and hybridized to phosphorothioated primers. HincII nicks its target site on the unmodified strand of the heteroduplex and exonuclease-deficient Klenow fragment then displaces the downstream DNA as it extends the 3'-end from the nick site. Amplification is exponential, and after the initial denaturation, is isothermal.

streak: The placement of a bacterial culture on solid media, especially with an inoculation loop or pipette, with the intention of covering the surface in a deliberate pattern.

stress-fibre: An actomyosin-containing filament of a non-muscle cell. Between mitoses (interphase) the fibres are anchored at membrane attachment sites and are involved in extension of lamellipodia and in cell motility.

striated muscle: Muscle that on microscopic examination shows a banding pattern that is due to the regular spacing of areas of overlap of thick and thin filaments within each sarcomere; includes skeletal muscle and cardiac muscle.

stringency: During nucleic acid hybridization or reassociation, the strictness with which the Watson-Crick base-pairing is required under specified conditions of temperature, pH, salt concentration, etc. Conditions of high stringency require all bases of one polynucleotide to be paired with complementary bases on the other; conditions of low stringency allow some bases to be unpaired.

stringent bacteria: (*see* magic spot nucleotides)

stroma: The supporting connective tissue of an organ. (*see also* parenchyma)

structural profile: An approach to prediction of similarities in protein conformations by comparison of amino acid sequences. A two- or three-dimensional plot of one or two features, e.g. optimal matching hydrophobicity, charge, likelihood of occurrence in -helix or -sheet, against their sequence number allows visual scanning to identify similarities of profiles and, presumably, similarities of conformations.

structure-activity relationship (SAR): An investigation into the dependence of a biological activity, e.g. pharmacologic or enzymatic, on the intimate details of chemical structure or conformation.

structure-activity relationships by mass spectrometry (SAR by MS): An approach to drug discovery. Sites of interaction of several small molecules with a macromolecule, e.g. aminoglycoside antibiotics with rRNA, are simultaneously probed by MS: the addition to each small molecule of a tag of characteristic mass, which does not affect its binding to the macromolecule, directs each complex to a different region of the mass spectrum. Analysis of fragments by tandem MS reveals the specific region of the macromolecule to which the small molecules bind.

structure-activity relationships by nuclear magnetic resonance (SAR by NMR): An approach to drug discovery. Pairs of low affinity molecules, each having an affinity constant in the micromolar range and binding to discrete areas of a biomolecule, as determined by NMR, are combined in various ways to happen upon a hybrid molecule that allows both ligands to attach and has a binding constant in the nanomolar range.

STS: (= sequence-tagged site (STS))

stutter band: (*see* slippage)

sub-: A prefix derived from the Latin for under or close.

subdomain model: (*see* framework model)

submitochondrial particle (SMP): A preparation obtained from mitochondria that is capable of electron transport.

subsite: (= specificity subsite)

substrate: A reactant in an enzymic reaction.

substrate anchoring: (*see* entropy effect)

substrate channelling: The sequential action of a multienzyme complex upon a substrate that prevents the dissociation of intermediate products.

substrate cycle: (= futile cycle (substrate cycle))

substrate inhibition: The inhibition of an enzymic reaction at high substrate levels; usually due to a second, inhibitory and lower-affinity binding site for the substrate in addition to the catalytic site.

substrate phage: A method for identification of the substrate specificity of a proteinase by its ability to select a phage that expresses a cleaveable polypeptide, which is one of a library of random amino acid sequences fused to a protein that tethers the phage to a solid support. Selection of the phage is by cleavage of the polypeptide. Several cycles of binding, proteolysis and phage propagation select the phage that expresses the scissile bond.

substrate-level phosphorylation: The metabolic synthesis of ATP that does not require the electron transport chain, e.g. the glyceraldehyde-3-phosphate dehydrogenase reaction.

subtractive DNA cloning: A difference cloning method; a technique for comparison of two closely related cell types (e.g. a cell with a deletion mutation and the normal cell), and isolation from one (the tester) of DNA that is absent from the other (the driver). Tester DNA is cleaved with a restriction nuclease that will allow its subsequent insertion into a cloning vector; an excess of driver DNA is randomly sheared. The two preparations are mixed, melted and annealed to form, among other species, duplexes composed of strands of the different sources (inter-resource duplex, IRD). The fraction of the annealed product, both strands of which were cleaved at restriction sites and which therefore is capable of being cloned, will be enriched in those sequences for which there was no competing randomly sheared driver DNA, i.e. those that are unique to the tester DNA, especially when annealing is terminated before an equilibrium is established (kinetic enrichment). The unique DNA sequences will therefore form a high proportion of the successfully cloned annealed DNA product. (*see also* magnet-assisted subtractive technique (MAST); subtractive hybridization)

subtractive hybridization: A variation of subtractive DNA cloning in which the restriction-nuclease-cleaved tester DNA is biotinylated before annealing with an excess of randomly sheared driver DNA. Annealing is halted at about 90% completion and the single-stranded DNA is separated from double-stranded DNA. As the unique tester fragments are present at lowest concentration, they will be slowest to find complementary strands and anneal and will, therefore, be enriched in the remaining single-stranded DNA fraction. They are isolated on the basis of biotin's affinity for immobilized avidin and subjected to subsequent cycles of purification.

subunit: One of the identical or non-identical protein molecules that make up a multimeric protein; also one of the ribonucleoprotein complexes that make up the ribosome. (*see also* heavy subunit; light subunit; protome)

sucrose-gradient centrifugation: A technique for characterization or preparation of subcellular particles. A swinging-bucket-type centrifuge tube is filled with a sucrose gradient, the bottom of which is most dense and the top least dense. A suspension of the particles is layered over the top of the solution, and centrifugation separates the particles within the gradient according to their density. At the termination of centrifugation the bottom of the tube is pierced and the sucrose, with particles of equal density, drips into a series of receiving tubes.

sugar: A simple carbohydrate that consists of a chain of carbon atoms, one of which is a carbonyl carbon (aldehyde or ketone), while the others bear hydroxy groups to give it the basic formula (CH2O)n, sometimes condensed into a hemiacetal or hemiketal; sugar derivatives include oxidation and reduction products, phosphate and sulphate esters, and amino derivatives.

suicide inhibitor: (*see* mechanism-based inhibitor)

sulfane: A persulfide, a compound of the type RS2H or RS2-; the group includes biosynthetic "reactive sulfur" compounds, which may be precursors of "inorganic"sulfur, e.g. non-haem iron proteins, and "organic" sulfur compounds , e.g. biotin. Sulfanes are operationally defined as a compound that reacts under alkaline conditions with CN- to form CNS-.

sulphatide: A sulphate ester of a ceramide or related compound, e.g. galactosyl-3-sulphate ceramide, ceramide-dihexoside sulphate.

sulphur cycle: The passage of sulphur from organic to inorganic forms, from substituents of cells to minerals, via the following steps: (1) uptake of sulphate from soil by organisms and conversion into organic sulphur, or reduction to the gas dimethylsulphide (e.g. by algae and cyanobacteria), (2) photochemical oxidation to methanesulphonic acid in the atmosphere, (3) return to the soil as it falls to earth in rain and snow, and (4) oxidation to sulphate by soil bacteria.

super-: A prefix derived from the Latin for above or beyond.

supercoiling: The torsional stress placed on double-stranded DNA, maintained for example by interaction with proteins or by DNA that is circular, and accommodated by a twist imposed on the duplex. A left-handed supercoil favours unwinding of the double helix; a right-handed supercoil favours tighter winding. (*see also* superhelical density (specific linking difference); topology)

superelectrophile: A chemical species with higher than usual electrophilic reactivity, often involving two positively charged centres. Superelectrophiles are intermediates in some enzymatic reactions; e.g. those of S-adenosyl methionine, in which the sulfonium group is juxtaposed with an acidic group on the adenosine moiety to promote sulfur-carbon bond cleavage and the consequent transfer of the methyl or aminopropyl group.

superfamily: A group of genes that are related by evolutionary divergence to an ancestral gene, but that code for products with different functions. (*see also* multigene family)

superhelical density (specific linking difference): A quantitative measure of the degree of supercoiling of a DNA molecule. The value, or , equals (L-Lo)/Lo, where Lo is the linking number of the relaxed double-stranded DNA molecule and L is the linking number for the supercoiled DNA. A negative value signifies a left-handed supercoil.

superhelix: The twisted axis of a fibrous polymer, such as double-stranded DNA, myosin tails or collagen. (*see also* coiled coil)

superinduction: The additional increase in the rate of synthesis of a protein caused by interference with the cellular apparatus for protein or mRNA synthesis that follows an increase in transcription of the gene. The mechanism varies with circumstances, but results in an increase in the steady-state level of the specific mRNA.

supernatant fluid: The unsedimented portion that remains after centrifugation. (*see also* pellet)

supersecondary structure: The arrangement of elements of secondary structures in a protein (a motif), e.g. -barrel, or in a nucleic acid, e.g. cloverleaf. (*see also* primary structure; quaternary structure; secondary structure; tertiary structure)

suppressor mutation: A mutation that reverses the effect of an earlier mutation, e.g. a mutation in a gene for a tRNA that permits it to read and override an amber mutation.

suppressor T-cell: (*see* T-cell)

supra-: A prefix derived from the Latin for above or beyond or prior to.

surface activation: (*see* esterase)

surface plasmon resonance (SPR): An optical technique for the detection and quantification of binding which has been incorporated into some biosensors. The material to which binding will be detected is immobilized on the surface of a very thin metallic film. When brought into contact with a solution of a ligand, the properties of the film are sufficiently altered so that light reflected from the other side of it will be detectably altered. In coupled plasmon-waveguide resonance (CPWR) spectroscopy, the properties of reflected polarized monochromatic light yield data on anisotropic membranes.

surface-enhanced laser desorption/ionization (SELDI): A pre-mass spectroscopy screening and selection technique. Spots on an inert surface are modified to bind a desired types of protein; e.g. with phenyl or amino groups for hydrophobic or acidic proteins, or with a specific ligands, such as a potential binding partners. After appropriate washings to remove unwanted proteins, the samples are desorbed and separated by mass spectroscopy to identify them by mass and fragmentation pattern.

surfactant: (= detergent (surfactant))

Svedberg unit: (*see* sedimentation constant)

switching: During development, the turning off of one gene and the turning on of a related one; e.g. during fetal development, the change from synthesis of -globin to synthesis of -globin.

syllabicity: (*see* zinc finger (metal-binding finger))

symmetry: One of the geometrical properties of a chemical or any other structure. Reflectional symmetry is the property of a molecule or object that permits it to be superimposed on its mirror image; rotational symmetry is a property that allows it to be rotated so that, in the new orientation, like groups exchange positions and the rotated orientation is indistinguishable from the original. (*see also* chirality; meso-carbon; Ogston hypothesis)

syn-: A prefix derived from the Greek for with.

syn conformation: (*see* Z-DNA)

synapomorphy: A characteristic common to a group originating from a common ancestor.

synapse: The structure at which a nervous impulse passes from one neuron, the afferent, to another, the efferent.

synaptic cleft: The narrow gap between two interacting neurons.

synaptic vesicle: A secretory vesicle that stores a neurotransmitter at the presynaptic membrane, i.e. on the afferent side of a synaptic cleft.

synaptosome: A vesicle formed by artificially pinching off and resealing the presynaptic ending of a neuron, i.e. the afferent portion of a synapse.

synchronous growth: The co-ordinated passage through the cell cycle of a large mass of cells; sometimes achieved by experimental starvation of the cells to bring them all to the start of S phase, then relieving the inhibition by refeeding.

syndecan: (*see* proteoglycan (mucopolysaccharide))

synkinesis: (*see* biosynthesis)

synkinon: (*see* biosynthesis)

synonymous: In molecular genetics, descriptive of a nucleotide substitution in a structural gene that does not result in an amino acid substitution; contrasted with non-synonymous, descriptive of base substitutions that do result in amino acid substitutions.

synonomous nucleotide substitution: = transparent mutation

synteny: The occurrence of closely similar blocks of DNA sequence in two different species. Synteny may be used as evidence of an evolutionary relationship, or as a basis for genomic mapping (*see*) *see* zoo-fluorescence in situ hybridization (zoo-FISH)

synthase: A lyase (acting in the reverse of the reaction) that adds one substrate across the double bond of another, e.g. -aminolaevulinic acid synthase, hydroxymethyl glutaryl-CoA synthase. (*see also* synthetase)

synthetase: An enzyme, regardless of its mechanism, that couples two substrates by a carbon-carbon, carbon-oxygen, carbon-sulphur or carbon-nitrogen bond that is driven by the hydrolysis of a phosphoanhydride bond, e.g. amino acyl-tRNA synthetase, fatty acyl-CoA synthetase. Those enzymes that catalyse such syntheses and do not use the energy of a phosphoanhydride bond are synthases (despite their frequently being misnamed).

synthetic peptide combinatorial library (SPCL): A strategy for determination of the specificity of a soluble peptide. Random collections of all the possible oligopeptides of a certain length, preceded by all the possible dipeptide sequences, are prepared; for example, if hexamers, there will be 400 (202) sets, each with the two known N-terminal residues followed by a random tetrapeptide; each of the 400 sets will consist of 160000 (204) hexapeptides. Each of the 400 is tested in an assay or bioassay, e.g. binding, inhibition, vasoconstriction. The successful set is refined by placing at the first of the random positions each of the amino acids, so there will be 20 sets of hexapeptides, each with a known tripeptide sequence followed by random tripeptide sequences. These are again tested in the assay system to further narrow the selection. Eventually the most favoured hexapeptide sequence is identified. A related strategy is positional scanning SPCL, in which a single amino acid is placed in each of the positions of the oligopeptide, a hexapeptide in the example, with each of the other five positions being filled at random by other amino acids. Each of these sub-libraries is assayed for selection and further refinement as in the original strategy.

synzyme: A synthetic enzyme, e.g. a polymer with an inert backbone, upon which dissociating groups like amines and carboxyls are mounted in a random fashion; or a protein product of in vitro evolution.

systematic evolution of ligands by exponential enrichment: (*see* cyclic amplification and selection of targets (CASTing))

T

-tome: A suffix derived from the Greek for cutting, which indicates a section or segment.

-troph: A suffix with the same derivation as tropho-.

"two-hit" hypothesis: The theory that some cancers, e.g. retinoblastoma, are the result of inactivating mutations in both alleles of a gene, each mutation being either inherited or somatic. Such a gene was originally termed an anti-oncogene, but more recently is refered to as a tumour suppressor.

T: series phage One of a group of bacteriophage that infect *E. coli*; the subject of extensive experimentation.

T phage: (*see* **bacteriophage**)

tachy-: A prefix derived from the Greek for fast.

TAD: Transactivation domain. (*see* trans-acting)

TA-DNA: A form of the DNA double helix found at TATA box sequences in complexes with binding proteins. TA-DNA differs substantially from A-DNA through a 50° greater tilt of the bases about the glycosidic bond. Insertion into B-DNA of TA-DNA of a number of base pairs fewer than the 11 to 12 required for a full turn of the TA-DNA helix results in a turn in the overall direction of the DNA chain, which helps adapt it to transcription activators and the up-stream RNA polymerase site.

tagged random PCR: A method for representatively sampling and amplifying DNA. The first two PCR cycles use primers of random sequence, and subsequent cycles use specific primers.

tailing: A procedure that is part of one method for insertion of a polynucleotide into a circular plasmid. Addition of a homopolymer to the 3'-end of one unit and the complementary homopolymer to the 3'-end of the other anticipates the annealing of these homopolymers and filling in of the gaps in the new duplex by a polymerase.

tailward: (*see* headward)

tandem array: A DNA structure in which a gene and associated sequences are repeated in an immediately adjacent position.

tandem enzymes: A pair of opposing enzymic activities that are resident in the same polypeptide chain, e.g. isocitrate dehydrogenase kinase and phosphatase; phosphofructokinase 2 and fructose bisphosphatase 2.

tandem mass spectrometry (tandem MS; MS/MS): An adaptation of MS for the analysis of molecular structure; an initial separation by MS generates a peak, i.e. a molecular ion or large fragment, that is further fragmented by gas-phase collision, the daughter fragments of which are themselves separated by MS. When more than two stages are involved, the technique is called *multi-dimensional MS* (MS^n, where *n* indicates the number of stages).

tandem MS: (= tandem mass spectrometry (tandem MS; MS/MS))

tannin: A product of plant origin that converts hides into leather; often polyphenols and aldehydes that cross-link collagen chains and render them insoluble and unreactive to enzymes that would degrade and putrefy untreated hides. Chromic acid salts are 'tanning agents' but not tannins.

Taq polymerase: *Thermus aquaticus* polymerase. (*see* polymerase chain reaction (PCR))

target cell: A cell that is stimulated by a particular hormone or neurotransmitter.

target detection assay: (*see* cyclic amplification and selection of targets (CASTing))

target organ: An organ that is stimulated by a particular hormone.

TATA box: A consensus sequence of eukaryotic DNA, named for its sequence on the coding strand, that aligns RNA polymerase II with the start site, about 25 bp downstream from it; from it; also known as a Hogness box or a Pribnow box. *see* TATA-less promoter

TATA-less promoter: A class of transcription promoter sequences that have no TATA box.

taut conformation: (*see* Monod-Wyman-Changeux (MWC) model)

tauto-: A prefix derived from the Greek for same, which indicates identity.

tautomer: An alternative arrangement of the chemical bonds of a molecule that requires movement of only electrons and protons, e.g. the enol form of a carbonyl.

taxon: A taxonomic group, such as a family or species.

TCA cycle: (= tricarboxylic acid cycle (TCA cycle)

T-cell: Thymus lymphocyte, which recognizes foreign proteins on the surface of other cells. The groups of T-cells include *cytotoxic T-cells* (*CTLs*; *killer T-cells*), which can kill the foreign cell; *helper T-cells*, which bind to and assist the proliferation of B-cells; and *suppressor T-cells*, which attenuate the action of B-cells. (*see also* anti-ergotypic; natural killer (NK) cell)

TDA: Target detection assay. (*see* cyclic amplification and selection of targets (CASTing))

Technical knock-out approach (TKO): Used for the construction of apoptosis-incapable mutations. Use of a cytokine to induce apoptosis in the vast majority of cultured cells allows the survival of only those that have deficient apoptotic apparatus.

TEF: (= thermic effect of food (TEF))

teichoic acid: A linear polymer present in Gram-positive bacterial capsules, cell walls and membranes, characterized by polyols (glycerol, ribitol, sugars), sometimes derivatized as D-alanine esters, and linked together through phosphodiester bonds.

telecrine function: The action of a hormone distant from its site of secretion. (*see also* paracrine)

telomere: The structure at the end of a eukaryotic chromosome formed by the DNA strand of the chromosome being extended by the enzyme telomerase, which is present in germ-line and cancer cells, and which adds a species-specific G-rich sequence. The DNA ends in a 3' overhang; this strand turns back onto itself, forming a telomere- (t-)loop and displaces its own up-stream sequence , forming a displacement- (D-)loop, and hybridizes with the complementary strand. The t- and D-loops are stabilized by telomere-binding proteins. The telomere, although not replicated by DNA polymerase, provides an anchorage for the RNA primer during replication. Without this extension, replication would shorten the chromosome each time that the cell divides. In fact this is exactly what happens in most cells, where telomerase is absent, the step-wise shortening of the chromosome creating a biological clock which counts down to eventual cell death.

temperate phage: A virus that infects a bacterium and inserts its genome into that of the bacterium, rather than lysing it.

temperature jump: (*see* relaxation)

template transfer: *see* bidirectional PCR amplification of specific alleles (bi-PASA)

template: A polynucleotide that encodes the information from which another polynucleotide, of complementary sequence, is synthesized.

tense form: (= taut conformation; see Monod-Wyman-Changeux (MWC) model)

ter: In enzyme kinetics, a designation of a reaction with four substrates or products; in genetics, a designation of the end of a chromosome. (*see also* centrosome; enzyme mechanism)

terminal glycosylation: The final modifications of the N-linked carbohydrate moieties of a protein that are effected in the Golgi apparatus. (*see also* core glycosylation)

terminal oligopyrimidine tract (TOP): 5'-TOP; an oligonucleotide sequence near the 5' cap site of a family of mRNAs that make them subject ot translational control.

termination factor: A protein that assists in the termination of the action of an RNA polymerase, e.g. the rho factor.

terminator codon: A polynucleotide triplet that signals the limit, at the C-terminus, of protein synthesis.

terpene: A compound synthesized from isoprene units (biosynthetically, isopentenylpyrophosphate) and recognized in products as a branched five-carbon motif. A *monoterpene* is composed of two isoprene units, a *sesquiterpene* of three, a *diterpene* of four, and a *triterpene* of six.

tertiary structure: The unique three-dimensional structure of a particular protein or nucleic acid. (*see also* primary structure; quaternary structure; secondary structure; supersecondary structure)

tester: (*see* subtractive DNA cloning)

tet-off: (*see* conditional expression)

tet-on: (*see* conditional expression)

tetra-antennary: Descriptive of a complex-type carbohydrate that has four unbranched oligosaccharide chains at a non-reducing end, often terminated by a sialic acid. Variants are *tri-antennary* and *bi-antennary* structures.

tetrad: In cytogenetics, the two pairs of chromosomes just after they have duplicated during mitosis and before they have separated. In the tetrad, the closely aligned strands have the opportunity for homologous recombination. On the molecular level, a tetrad is a four-stranded DNA structure, e.g. the i-tetrad, composed of parallel and anti-parallel poly(C) sequences in which all strands are equivalent and each C-C base pair shares a proton.

tetrahedral intermediate: A proposed transition state in the hydrolysis of a peptide or ester bond in which a nucleophile, water or a functional group of the enzyme catalytic site has added across the carbonyl group.

T-form: Tense or taut form. (*see* Monod-Wyman-Changeux (MWC) model)

thanatogene: (*see* oncogene)

thermic effect of food (TEF): The energy expended on digestion of food. It is also known as specific dynamic action.

thermo-: A prefix from the Greek for heat.

thermodynamic model of protein structure: (*see* framework model)

thermogenesis model: A theory of the origin of life, particularly of how prebiotic structures utilized environmental energy to drive synthetic processes. Basing itself on the current belief that the most primitive organisms are the thermophilic micro-organisms, the theory holds that prebiotic synthesis was driven by temperature variation, perhaps owing to convection currents near undersea thermal vents. A population of proteins is assumed, having been formed by processes postulated by other theories of the origin of life. Among these proteins was the original enzyme, an ATP synthase, which underwent a reversible temperature-dependent folding/unfolding: cycles of temperature variation drove the synthesis of ATP, or other phosphoanhydride, by binding substrates at a lower temperature and facilitating synthesis of an anhydride bond on the protein surface; as the protein unfolded at the higher temperature, it dissociated the ATP (protein-associated thermosynthesis). When the components of membranes became available, the synthase activity was incorporated into them and formed primitive chemosmotic vesicles (membrane-associated thermosynthesis). Variations in incident light e.g. under the waves of shallow water, might have energized primitive photosynthetic vesicles.

thermogenesis: The process of generation of heat by the uncoupling of electron transport, especially in brown adipose tissue.

thermophile: (*see* extremophile)

thick filament: In muscle cells, an aggregate of myosin molecules in which the long, fibrous tails are intertwined in a superhelix that leaves the globular heads as knobs in a helical array over the surface. The myosin molecules are oriented so that their tails are directed towards the centre of the filament and consequently leave this region of the filaments bare of the knobby projections.

thin filament: In muscle and other cells, polymerized actin and associated proteins

thin-layer chromatography (TLC): A technique to separate small molecules on a thin film of an adsorbent on a glass or semi-rigid plastic surface by their differential mobilities in a liquid phase that passes through the film by capillary action.

thio-: A prefix derived from the Greek for sulfur.

thiocyanate degradation: A method for analysis of peptide structure by repetitive sequential degradation from the C-terminus. The free a-carboxyl group is activated with acetic anhydride and reacted with thiocyanate, and the terminal residue is then hydrolysed off as a thiohydantoin. (*see also* Edman degradation)

thiol proteinase: (= cysteine proteinase (thiol proteinase))

thiolysis: In β-oxidation, cleavage of the bond between the β- and γ-carbons by coenzyme A to form acetyl-coenzyme A from carbons α and β and a coenzyme A thioester from the alkyl-carbonyl moiety, i.e. carbons originally γ, δ, ...

thiostatin (T-kininogen): An α_1-cysteine proteinase inhibitor and kininogen of the rat.

thiosulphate shunt: A pathway for anaerobic metabolism of sulphide by bacteria, e.g. in marine sediments; thiosulphate (S_2O32^-) is an intermediate between sulphide (S^{2-}) and sulphate (SO_42^-) in reduction to the former, oxidation to the latter and disproportionation to both.

thiotemplate mechanism: A non-ribosomal, non-RNA-dependent mechanism for the synthesis of bacterial cyclic oligopeptide antibiotics (e.g. gramicidin S) in which the amino acid residues are first transferred from amino acid adenylates to thiol groups of a peptide synthetase that then collects the residues, in sequence, on a phosphopantetheine arm until the mature peptide is cleaved off by a cyclizing transpeptidation reaction.

thromboxane: An eicosanoid that features a six-membered cyclic ether nucleus.

thrombus: The product of coagulation, composed of cells and proteins, found when blood clots.

thylakoid membrane: A membranous structure within chloroplasts that harvests light and synthesizes ATP.

thymineless death (TLD): The lethality of thymine starvation in thymine auxotrophs. The phenomenon is believed to be due to an inability to repair double-strand breaks in DNA.

tier: The cohort of chains of a branched-chain polysaccharide that all have the same relationship to the branch points. The outer tier of glycogen is all those glucosyl units that are attached in 14 bonds that run from the non-reducing ends to the first 16 branch points; the second tier is those glycosyl units between the first and second outer branch points, etc. (*see also* limit dextrin)

tight junction: The seal that closes the gap between adjacent epithelial cells to ensure closure of a vascular space. Transport through a tight junction is *paracellular*, in contrast to *transcellular* transport through cells.

time constant: In first-order kinetics, the reciprocal of the rate constant.

time-of-flight mass spectrometry (TOF-MS): A variant of MS in which the mass of a molecular ion is calculated from the time of travel from the point of ionization to the detector. TOF-MS uses a pulsed ionizing source such as a laser or ^{252}Cf decay into one daughter isotope that ionizes the test material and a second that signals the beginning of the pulse. (*see also* mass spectrometry (MS))

time-resolved crystallography: (= Laue crystallography (time-resolved crystallography))

time-resolved: Decribes a technique that follows a parameter over the time in which it changes, e.g. time-resolved fluorescence applied to monitor changes in protein folding; as opposed to a technique that observes only the initial and final states.

TIMP: (= tissue inhibitor of metalloproteinse)

tissue engineering: A group of procedures for replacement of damaged or non-functional cells or organs with homologously or heterologously transplanted cells.

tissue inhibitor of metalloproteinse: TIMP; matrixin; a natural inhibitor of the metalloproteinases that remodel structural proteins of connective tissues.

tissue: A group of cells, of one or several types, that serve a specific function, e.g. adipose tissue, epithelium.

titin: An exceptionally large muscle protein, 3000kDa in molecular mass and 1m in length, that spans the entire length of a sarcomere.

titration curve: A graphical representation of the protonic dissociation of a compound, e.g. pH against equivalents of alkali added to the acid form of the compound.

T-jump: Temperature jump. (*see* relaxation)

TLC: (= thin-layer chromatography (TLC))

t-loop: (*see* telomere)

T-lymphocyte: (= T-cell)

TOF-MS: (= time-of-flight mass spectrometry (TOF-MS))

tolerance: The property of the immune system that prevents development of autoimmune responses; the suppression of lymphocytes that would otherwise produce autoantibodies. (*see also* activation)

tomo-: A prefix with the same derivation as *-tome.*

topo-: A prefix derived from the Greek for place.

topoisomers: Double-stranded DNA molecules that differ only by their linking numbers.

topology: The nature of the supercoiling of a double-stranded DNA molecule; also, in protein chemistry, a formalized array of secondary structures within a molecule. (*see also* Greek key)

topology/packing diagram: A schematic representation of secondary structure of a protein in which -helix is rendered as a circle, -sheet as a rectangle and connecting sequences as lines. The circles and rectangles are arranged in two dimensions to approximate their true orientation in space. Proteins fall into four classes: all-, which have only helix secondary structure; all-, which have only sheet; +, which have regions of both helix and sheet; and /, which have approximately alternating helix and sheet. (*see also* Rossmann fold)

toroidal: (*see* plectonemic)

torsional stress: (*see* entropy effect)

totipotency: The capacity of a zygote to develop into a fully differentiated organism, a property lost as the zygote develops into more specialized cells that are more committed to particular lines of development.

toxin: A natural poison, especially one of a class of proteins that act intracellularly to interrupt vital functions. *see also* endotoxin; *see also* exotoxin; *see also* ribosome-inactivating protein (RIP)

toxoid: A toxin which has been treated, e.g. with formaldehyde, to inactivate its pathogenicity but not its immunogenicity.

TPN+: Triphosphopyridine nucleotide; obsolete name for $NADP^+$.

TPNH: Reduced triphosphopyridine nucleotide; obsolete name for NADPH.

trace element: A constituent of a tissue, a cell or the diet that is present in very low amounts, e.g. copper, zinc, molybdenum.

tracer: A compound that is labelled by a radioactive or non-radioactive isotope to assist in the study of its transport or metabolism.

traffic ATPase: A class of transmembrane transporters that includes the cystic fibrosis transmembrane regulator (CFTR) and transporters responsible for multidrug resistance. Transport is coupled to ATP hydrolysis but not to the counter- or co-transport of any other metabolite.

trafficking: The transport of substances, e.g. a newly synthesized or modified nucleic acid or protein, from one intracellular structure to another. Intracellular transport of proteins is known as protein kinesis.

trancytosis: (*see* caveolae)

trans: (*see* cis)

trans-acting: Descriptive of a controlling effect of a regulatory gene on a structural gene at some distance from it on the same chromosome or on a different chromosome. The terminology was later broadened to distinguish intermolecular (*trans*) from intramolecular (*cis*) actions. A regulatory factor mediates binding of activators and repressors via its DNA-binding domain and its transactivation domain. At higher levels, the concentration of the factor exceeds that of its DNA sites and, in solution, it can bind the activator or repressor so as to preclude binding of the latter to any DNA-bound factor, and can thus prevent or 'squelch' activation or repression. (*see also* cis-acting)

transamidation: The substitution of one amino compound for another in an amide linkage, e.g. the cross-linking of fibrin by the displacement of NH_3 from glutamine residues by an -amino group of a lysine residue to form an isopeptide bond.

transamination: The concurrent amination and reduction of a carbonyl of one compound as another is deaminated and oxidized; especially the conversion of one -amino acid into its corresponding -oxo acid as another -oxo acid is converted into its corresponding -amino acid, mediated by the pyridoxal phosphate prosthetic group of a transaminase (aminotransferase). (*see also* quinimine form)

transcribed strand: (*see* antisense)

transcription: Synthesis of RNA from a DNA template.

transcription factor: A protein that binds to a specific DNA sequence and enhances or suppresses transcription of the target gene. One class is the basic helix-loop-helix proteins; upon associating with another molecule through their helix-loop-helix regions, the basic sequences of the dimer binds the target DNA.

transcriptional activation: The process of separation of strands of DNA at which replication will commence. Short RNA sequences hold apart the DNA strands to allow a primosome to bind and synthesize primers of DNA synthesis.

transcriptional decoy model: (*see* insulator)

transcriptional silencing: The exercise of genetic control to prevent expression of a structural gene.

transcription-translation coupling: The immediate use of mRNA transcripts for protein synthesis before the synthesis of the mRNA is complete; occurs in bacteria where RNA processing does not occur.

transcriptome: A group genes that are transcribed under a certain set of conditions.

transcriptomics: *see* (omic research)

transcytosis: The transport of materials across a polarized cell, e.g. in digestion from the apical surface of a cell of the intestinal epithelium and its movement across the cell to the basolateral surface.

trans-dominant: Descriptive of an engineered protein or nucleic acid fragment that inhibits the activity of its parent molecule by preferential binding to the parent's target site.

transduction: The conversion of one kind of energy into another, e.g. conversion of the chemical energy of ATP into mechanical energy by muscle contraction; also conversion of a signal as it crosses a barrier, e.g. linkage of hormone binding on the outside of a plasma membrane to the generation of a second messenger inside it; also DNA transfer by a virus that can incorporate into its own genome part of the DNA of a first host and then transfer it to a second host. *see* conjugation; transformation

transesterification: The substitution of one alcohol for another in an ester bond, e.g. the displacement of one 3'-hydroxy group of a phosphodiester bond by the 3'-hydroxy group of another nucleotide during the self-splicing of an RNA molecule, or the internal transfer of phosphate from a 3'-hydroxy to a 2'-hydroxy group during the reaction of pancreatic ribonuclease.

transfection: The process by which viral or bacteriophage DNA is introduced into a cell or bacterium. (*see also* transformation)

transfer RNA: (= tRNA)

transfer vesicle: A portion of the endoplasmic reticulum which encloses newly synthesized proteins, that buds off and travels to and fuses with the Golgi apparatus; also those vesicles that transfer protein from the Golgi apparatus to lysosomes, secretory vesicles and the plasma membrane.

transferase: One of a class of enzymes that transfer a chemical group from donor substrates to acceptor substrates, e.g. a kinase, a phosphorylase, a transaminase. (*see also* hydrolase)

transformant: A product of transformation; a transformed cell.

transformation: The process by which a cell line, that can normally be expected to undergo a limited number of cell divisions before death, becomes immortal; also the process by which isolated foreign DNA is introduced into a cell or bacterium.

transforming factor: In the classical experiments by Avery, MacLeod and McCarty, the name given to the agent that caused the heritable transformation of R strains of *Streptococcus pneumoniae* into S strains. The identification of the factor as DNA was the definitive proof that DNA is the repository of genetic information. (*see also* lipopolysaccharide (LPS))

trans-fusion: (*see* alpha (α)-configuration)

transgene: DNA that has been experimentally introduced into a transgenic animal.

transgenic: Descriptive of an animal that has developed in a surrogate mother from a fertilized ovum that had been injected with a recombinant DNA gene; contrasted with *congenic*, meaning of the same genetic origin.

trans-Golgi: (*see* Golgi apparatus)

transition state: In chemical kinetics, the hypothetical state that is mid-way between reactants and products, poised at a point where the reaction is as likely to go forward to products as it is to fall back to reactants. If the reactants and transition state may be considered to be in equilibrium, application of the second law of thermodynamics allows analysis of the temperature-dependence of reaction rates. The concept may be applied to enzyme kinetics to describe the transition from enzyme-substrate complex to enzyme-product complex.

transition state inhibitor: An enzyme inhibitor that is designed to fit the transition state of an enzyme, as opposed to one that is a substrate analogue; characterized by a very low dissociation constant.

transition: In replication or transcription, an error in which one purine is substituted for another, or one pyrimidine for another. (*see also* transversion)

transition temperature: The temperature at which the plasma membrane undergoes a phase transition, due to the increased mobility at higher temperatures of the fatty acyl chains of phospholipids.

translation: The synthesis of a protein directed by mRNA.

translocation: In genetics, the movement of a portion of one chromosome to another; in protein synthesis, the transfer of the newly elongated peptidyl-tRNA from the amino acyl site to the peptide site of a ribosome; in cell biology, the movement of a molecule across a barrier or between cytosol and membrane surface.

transmembrane domain: A feature of most intrinsic proteins of plasma or vesicular membranes; a polypeptide sequence of about seven residues if -sheet, up to 22 residues if -helix, that connects extracellular to intracellular domains, joined by extended polypeptides on the cytoplasmic and external or vesicular sides. Receptor and ion channels have between one and twelve such domains.

transmethylation: The metabolic transfer of preformed methyl groups from one acceptor to another, e.g. from *S*-adenosylmethionine to guanidinoacetate to form creatine.

transparent mutation: A single base change that does not result in an altered protein, due to the degeneracy of the genetic code, i.e. the mutation is from one codon to another for the same amino acid. (*see also* silent mutation)

transpeptidation: A characteristic of some proteinases; the transfer of one product of the catalytic cleavage to another peptide rather than to water; the transfer of the N-terminal fragment to the amino group of an acceptor peptide or the transfer of the C-terminal fragment to the carboxyl group of an acceptor; the former often, but not exclusively, due to an acyl-enzyme intermediate in the catalytic cycle.

transposable element: (= transposon)

transposition: The movement of a mobile element into or out of a chromosome or plasmid.

transposon: One of a class of genes that are capable of moving spontaneously from one chromosome to another, or from one position to another in the same chromosome; also known as a jumping gene or a transposable element. Transposons are found in the genome flanked by inverted terminal repeats.

transposon tagging: The use of a transposon to isolate a gene whose product may be unknown. Mutants in which a transposon inserts into and disrupts the function of a gene are selected by phenotype.

A DNA fragment that contains the gene with its insertion is recognizable and is selected by its binding of a probe complementary to the transposon insert. The DNA whose sequence flanks the insert is isolated and used as a probe to screen a library of non-mutants for the intact gene.

trans-recognition: A refinement of the N-end rule; the determination of the rate of degradation of a cellular multimeric protein composed of different kinds of subunits by amino acid residues of a subunit other than the one that is rapidly degraded. *cis*-Recognition is the targeting by residues of the subunit that is rapidly degraded. The distinction implicitly acknowledges that such subunits may be degraded at different rates.

trans-splicing: Discontinuous transcription; the formation of a chimeric mRNA from two transcripts of unlinked loci. *trans*-Splicing, as distinct from *cis*-splicing, the more familiar process of excision of introns from pre-mRNA, is common in trypanosomes. An untranslated sequence at the 5′ end of the mature trypanosome mRNA is a transcript of a tandemly repeated mini-exon and is termed mini-exon-derived RNA (medRNA). It is ligated to a transcript of the coding sequence by displacement of a mini-intron (minRNA) at the 5′ end in a process mechanistically akin that of *cis*-splicing.

transverse diffusion: (= flip-flop (transverse diffusion))

transverse relaxation-optimized spectroscopy (TROSY): A variation on correlation spectroscopy (COSY), the two-dimensional NMR technique for correlation of the chemical shifts of two paramagnetic nuclei, which gives much sharper linewidths than COSY and which thus extends the usefulness of the method to much larger macromolecules. Decreased linewidth is achieved by omission of the decoupling step of COSY, in which the four peaks which arise from ^{15}N-^{1}H or ^{13}C-^{1}H interactions are combined to preserve their intensities at the cost of sharpness of linewidth; in TROSY, only the strongest peak of the four is selected.

transversion: In replication or transcription, an error in which a purine is substituted for a pyrimidine, or a pyrimidine for a purine. (*see also* transition)

transwinding: The action of a protein to separate one RNA strand from its hybrid with a second, and to promote its reannealing with a third RNA strand.

treadmilling: The dynamic state of microtubules, the addition of subunits at one end, and their removal from the opposite end, which may account for the movement of microtubules in the cell. The nucleation site, e.g. at the centromere of a chromosome, the minus end of a microtubule cannot exchange subunits until it is released, which happens frequently, to allow treadmilling to proceed.

trefoil: (*see* node)

tri-: A prefix derived from the Greek and Latin for three.

tri-antennary: (*see* tetra-antennary)

tricarboxylic acid cycle (TCA cycle): The metabolic pathway in which acetyl-CoA is catalytically oxidized to carbon dioxide, with the concomitant reduction of NAD^+ and FAD via a series of tricarboxylic (citric, cisaconitic and isocitric) and dicarboxylic (succinic, fumaric, malic and oxaloacetic) acids. The actual carbon atoms that appear as CO_2 after a single passage through the cycle are not identical to those that entered as acetyl-CoA. Also known as the citric acid cycle or the Krebs cycle.

trigonal bipyramid: (*see* pentacovalent intermediate)

trinucleotide repeat: A mutation in which a trinucleotide is repeated, even thousands of times; a characteristic of several diseases, e.g. fragile X syndrome.

tripartite leader: In late viral mRNAs, the 5'-non-coding region; an extended polynucleotide sequence characterized by strong secondary structure between sequences of little structure. (*see* ribosome jumping)

triple helix: The tertiary structure of collagen that twists three polypeptide chains around themselves; also a triple-stranded DNA structure that involves Hoogstein base pairing between B-DNA and a third DNA strand that occupies the major groove.

triplet: A three-base codon of the genetic code. Also a component of the mitotic spindle, seen in cross-section as one of nine structures, each composed of a microtubule doublet with an additional series of protofilaments that constitute a third subunit; the nine triplets radiate from a central axis and appear as a *cartwheel*.

triplet state: An excited electronic state in which an electron is raised to a higher energy level and its spin is reversed, which prevents its easy collapse to the ground state. Such collapse is sometimes accompanied by phosphorescence. (*see also* singlet state)

triplex DNA: Triple-stranded DNA in which the third oligonucleotide strand lies in the major groove of duplex DNA. This strand may also be a synthetic oligonucleotide with a sequence designed to target a specific sequence of a duplex; it can block transcription or, if it bears a covalently-attached reactive group, it can cleave the duplex at a unique site. (*see* Hoogstein base pairs)

triplex-forming oligonucleotide (TFO): An oligonucleotide that recognizes a specific dsDNA sequence. Designed to affect the expression of a specific gene, it lies in the major groove of the double helix and forms Hoogstein base pairs with the purine-rich strand.

tris-: A prefix derived from the Greek for thrice.

triskelion: A Manx key, a heraldic emblem that depicts three legs seen bent and in profile which radiate from a common centre; by analogy, therefore, the three-legged structure of clathrin which is composed of three heavy and three light chains. Subunits of the polyhedral clathrin shell surround coated vesicles.

triterpene: (*see* terpene)

tRNA: Transfer RNA; the RNA that serves in protein synthesis as an interface between mRNA and amino acids. It carries an anticodon sequence that pairs bases with a codon of mRNA, and it binds an amino acid at its 3'-end through an ester bond.

Trojan horse inhibitor: Affinity labelling reagent.

trophic: Descriptive of the dependence of a cell, tissue or organism on an external agent for its maintenance or growth. (*see also* tropic)

tropho-: A prefix derived from the Greek for nourishment, which indicates dependency.

tropic: Descriptive of a response of a cell, tissue or organism to a stimulus, especially a hormone. (*see also* trophic)

tropo-: A prefix derived from the Greek for turn which indicates attraction.

trp operon: A region of DNA that, when expressed, results in an RNA transcript that codes for five enzymes that act sequentially on chorismic acid to transform it into tryptophan.

truncation: Elimination of the N- or C-terminal portion of a protein by proteolysis or manipulation of the structural gene, or premature termination of protein elongation due to the presence of a termination codon in its structural gene as a result of a nonsense mutation.

ts: A prefix that indicates a temperature-sensitive form.

tumbling: The irregular motion of a flagellar bacterium in solution. (*see also* random walk)

tumour initiator: An agent that damages cellular DNA, a necessary condition for tumorigenesis. (*see also* anti-oncogene (tumour suppressor gene); tumour promoter)

tumour promoter: An agent that converts a cell with damaged DNA into a malignant cell. (*see also* anti-oncogene (tumour suppressor gene); tumour initiator)

tumour suppressor gene: (= anti-oncogene (tumour suppressor gene))

tumour suppressor: A product of the expression of an anti-oncogene.

tumour: A neoplasm; a growth of new tissue, sometimes malignant.

TUNEL technique: The method of terminal deoxy-transferase (TdT) and dUTP assisted nick end labeling for detection of apoptosis in single cells. The method takes advantage of the fact that the DNA of apoptotic cells is extensively nicked. With TdT, biotinylated poly-U sequences are introduced at the nick sites and then conjugated with avidin bound to peroxidase. Peroxidase is used in an *in situ* chromogenic assay. The method is applicable to histological sections and to populations of cells, which can then be separated and analysed by flow cytometry.

turgor: The internal hydrostatic pressure of a cell, especially a plant cell.

turnover number (catalytic-centre activity): The number of substrate molecules that a single catalytic site of an enzyme molecule can convert into product in a given time. (*see also* catalytic rate constant (k_{cat}))

turnover: Degradation and resynthesis of a compound or macromolecule.

twintron: An intron within an intron; the internal intron is excised before the external intron is spliced out.

twist conformation: A conformation of a five-membered ring in which no four centres lie in the same plane. (*see also* chair form; envelope conformation)

twisting number (T): A topological property of double-stranded DNA; the number of turns one DNA strand makes around the other; the number of paired bases in a DNA molecule divided by 10; equal to the linking number minus the writhing number ($T = L - W$).

two-component pathway: A pattern of molecular organization by which responses to external stimuli are processed by cells to effect modulation of output, such as chemotaxis in bacteria, or osmolarity regulation in plant cells. An example is a kinase that, following its interaction with an activated receptor, phosphorylates a function-specific response regulator.

two-dimensional DNA typing: Characterization of DNA according to base composition by electrophoresis, first according to chain length in one direction, then at 908 in a denaturant (e.g. formamide, urea).

two-dimensional electrophoresis: Electrophoretic separation of proteins on a solid support by one technique in one dimension, then by another technique in another. For example, proteins are subjected to isoelectric focusing in a long, thin tubular gel, then the developed gel is placed at the top of a sodium dodecyl sulphate/ polyacrylamide slab and subjected to gel electrophoresis in the other direction.

two-dimensional homochromatography: A form of displacement chromatography in which, after an initial separation in the first dimension of a labelled oligomer (e.g. an oligoribonucleotide), separation in the second dimension proceeds in the presence of a high concentration of a heterogeneous mixture of oligomers (e.g. an unfractionated endonuclease digest of yeast RNA).

two-hybrid system: An assay for detecting *in vivo* protein-protein interactions, which depends upon expression of a reporter gene, usually in yeast. As transcription of eukaryotic enzymes are regulated by transcriptional activators having distinct DNA-binding and transcriptional activator domains, the coding sequences for each of these domains can be fused with coding sequences for each of the two proteins to be tested; if the proteins bind, the DNA-binding and transcriptional activator domains will be reunited, permitting transcription of the reporter gene. Typically, the system is used to detect a binding partner of a known protein (encoded in the bait vector as a chimera with the DNA binding domain of the transcription factor) to a candidate partner (encoded in the prey vector as a chimera with the activation domain of the transcription factor), which is the homologue of a known binding partner or is encoded by a cDNA library.

two-state hypothesis: In protein chemistry, the postulate that the folding of a protein occurs without any stable intermediate between the denatured and native states. (*see also* framework model; molten globule intermediate; funnel concept)

Tyndall effect: The phenomenon of light scattering that is dependent upon the wavelength of light. The intensity of light scattered by small isotropic particles is inversely proportional to the fourth power of the wavelength.

type I marker: When assigning landmarks for genome mapping, a type I marker is an expressed gene. A polymorphic microsatellite sequence is a type II marker.

type II marker: (*see* type I marker)

tyrosine kinase: An enzymic activity associated with the cytoplasmic domain of several growth factor receptors. Two such molecules, when liganded to an effector molecule by their extracellular domains, phosphorylate tyrosine residues on each other; this increases their kinase activities so that they are then able to phosphorylate other cytoplasmic proteins, thus continuing the signal transduction pathway initiated by effector binding to the receptor.

tyrphostin: An inhibitor of a tyrosine kinase.

U

U-DNA: (*see* Kunkel mutagenesis (dUTP system mutagenesis))

ubiquitin pathway: A route to cellular proteolysis that depends upon conjugation of the targeted protein to a small protein, ubiquitin, followed by hydrolysis.

ultrafiltration: A separation procedure in which a solution is forced through a membrane with a pore size that is selected to retain macromolecules of a certain size and to pass smaller ones.

ultraviolet (UV) spectroscopy : A technique which measures absorption of electromagnetic radiation at wavelengths shorter than the visible spectrum, i.e. below 400 nm. In biological samples, UV spectroscopy detects aromatic residues of proteins and bases of nucleic acids. Commercial UV spectrometers commonly also measure the visible spectrum, i.e. up to 780 nm, and detect extensively conjugated aromatic systems, such as haemoglobin and chlorophyll, but also transition metals and their complexes. Spectra are commonly presented as absorption or molar extinction coefficient vs. wavelength.

unassigned reading frame (URF): An open reading frame for which a protein product has not yet been identified.

uncompetitive inhibition: A form of enzyme inhibition in which the inhibitor binds to the enzyme-substrate complex, resulting in decreases in the K_m and V_{max} values. (*see also* competitive inhibition; inhibitor; non-competitive inhibition (mixed inhibition))

uncoupling agent: A compound that dissociates electron transport from ATP synthesis and allows transport to proceed without synthesis. (*see also* oxidative phosphorylation)

unfolded protein response (UFR): The cell's graded response to the presence of unfolded or misfolded proteins; initially, activation of the genes that support endoplasmic reticulum-associated protein degradation, then, if the system is incapable of coping with the magnitude of the challenge, apoptosis to sacrifice the entire cell.

uni: (*see* enzyme mechanism)

uniport: A transport mechanism that drives a single compound or ion across a membrane, not coupled with transport of any other compound or ion. (*see also* antiport; mobile barrier; mobile carrier)

unit cell: The basic repeating unit of a crystal.

unit evolutionary period: A measure of the rate of divergence of two daughter genes after duplication of the ancestral gene; the time taken to accumulate 1% divergence in primary structures.

unit membrane Danielli-Davson model: A model for the structure of biological membranes; a variation of the Danielli-Davson model in which the membrane is composed of plaques of bilayer (the unit membrane); the outer leaflet turns over the edges of the plaque to become continuous with the inner leaflet. (*see also* fluid mosaic model (Singer-Nicolson model); Gortner and Grendel model)

unit: In enzymology, a measure of enzyme activity; usually the conversion of 1mol of substrate per min under specified conditions. (*see also* katal)

universal: In molecular biology, descriptive of a tool that is generally useful, not limited to a specific DNA fragment or sequence; e.g. a universal adaptor is one that can be ligated to all DNA fragments.

unnatural amino acid mutagenesis: A method for replacement of an amino acid residue of a protein with a non-naturally occurring amino acid. The mRNA that encodes the protein is modified to encode a nonsense suppressor codon. For example, mRNA is modified to use the UAG codon; this mRNA, together with a nonsense suppressor tRNA with the anticodon CUA, and with an unnatural amino acid at its 3'-end, is introduced into an expression system, possibly a *Xenopus* oocyte.

unscheduled DNA synthesis: Replication that occurs outside the synthetic (S) phase of the cell cycle.

untranslated region (UTR): A genomic DNA sequence that is not translated into an RNA sequence.

upfield: Descriptive of a resonance position in a magnetic field higher than that at which a standard displays its signal.

upstream: In a polynucleotide chain, towards the 5'-end. (*see also* downstream)

urea cycle (Krebs urea cycle): The metabolic pathway that receives nitrogen and carbon dioxide, as carbamoyl phosphate and aspartic acid, and synthesizes urea.

ureo-: A prefix derived from the Latin for urine.

ureotelic: Descriptive of an organism which synthesizes and excretes urea as a final product of nitrogen metabolism, such metabolism and excretion being ureotely. (*see also* ammonotelic; uricotelic)

URF: (= unassigned reading frame (URF))

uricotelic: Descriptive of an organism which synthesizes and excretes uric acid as a final product of nitrogen metabolism, such metabolism and excretion being uricotely. (*see also* ammonotelic; ureotelic)

uronic acid: A sugar derivative in which the hydroxymethyl group is oxidized to a carboxyl group.

UTR: (= untranslated region (UTR))

V

V(D)J recombination: The process of assembling the DNA sequences that encode an immunoglobulin or a T-cell receptor (TCR), the cell-surface protein that recognizes an antigen. The process excises polynucleotide sequences between V (variable), D (diversity) and J (joining) sequences, with the assembly of a coding sequence that will encode a complete potential immunoglobulin or TCR. The mechanism involves excision at recombination signal sequences (RSSs) that are characterized at each end of the to-be-excised sequence by 23 and 12 bp sequences (the 12/23 rule) that will be fused together in the signal joint as the excised sequence is cyclized. The ends of the coding sequences, which were sealed as hairpin turns during the first phase of the splicing, are reopened, an overhang is filled in with the insertion of so-called P nucleotides, and ligated to form a coding joint. (*see* double-strand-break repair model)

v: A prefix that denotes viral. (*see also* oncogene)

vaccinome: A plasmid that contains a DNA sequence which encodes an immunologically effective protein, such as an HLA T-cell epitope. A string-of-beads vaccine plasmid is a vaccinome that contains many such sequences.

vaccinomics: *see* omic research

vacuole: A large vesicle, especially in a plant cell.

van den Bergh reaction: A method for colorimetric estimation of serum bilirubin by coupling with diazotized sulphanilic acid. The *direct van den Bergh reaction*, in aqueous medium, yields the 'direct' bilirubin, i.e. the amount of water-soluble bilirubin glucuronide conjugates. Upon prior addition of methanol to solubilize free bilirubin, the total bilirubin is found, and by this *indirect van den Bergh reaction* the 'indirect' bilirubin, i.e. the amount of unconjugated bilirubin, may be calculated as the difference between the total and direct levels.

van der Waals bond: The weak attraction of neighbouring neutral atoms that includes dipole-dipole and dipole-induced dipole interactions and London dispersion forces.

van der Waals radius: The dimensions of an atom deduced from its packing in crystals.

Van Slyke method: A method to quantify primary amino groups by measurement of the volume of nitrogen produced upon reaction with nitrous acid; an obsolete method for the assay of proteinase action.

variable number tandem repeat (VNTR): A minisatellite DNA; the repetition of a 35-80bp sequence, to a size of up to 2kbp, that is characteristic of an individual. VNTRs are used in forensic science to identify, or exclude, suspects of a crime. (*see also* DNA fingerprinting (DNA profiling))

variable region: The sequences of immunoglobulin light or heavy chains that show variation from one particular antibody to another, and are responsible for the binding of antigen. (*see also* constant region; hypervariable region)

vaso-: A prefix derived from the Latin for vessel, which indicates the structures that contain and conduct blood and other body fluids.

vasoconstriction: The decrease in diameter of small blood vessels, especially in the kidney, that raises blood pressure. (*see also* **vasodilation**)

vasodilation: The increase in diameter of small blood vessels, especially in the kidney, that lowers blood pressure. (*see also* vasoconstriction)

VCD: Vibrational circular dichroism. (*see* vibrational optical activity (VOA))

vector: An animal host and carrier of a pathogen; e.g. the flea which transmits the bacterium *Pasteurella pestis*; also, a DNA molecule that can be replicated in a cell and that can serve as the vehicle for transfer to such a cell of DNA that has been inserted into it by recombinant techniques.

vectorette PCR: A PCR variation that permits specific amplification of the sequence from a single known internal sequence to the end of the fragment. The source DNA is cleaved by a restriction endonuclease that leaves an overhang, by which the fragments are ligated to a compatible universal synthetic oligonucleotide duplex called a vectorette. The vectorette is engineered with a central mismatched sequence, and the vectorette primer is identical to the sequence on the vectorette strand that becomes ligated to that fragment strand which can bind a second primer that is complementary to the known internal sequence of the fragment. The binding site for the vectorette primer is created only when the internal primer is extended to the end of the vectorette. This avoids duplication of the vectorettes attached to all other restriction fragments. A refinement is *splinkerette PCR*, in which the primers are designed to form hairpin structures, which decreases the danger of non-specific priming. (*see also* inverse PCR)

Velcro mechanism (cysteine switch): An activation mechanism for matrix metalloproteinase family zymogens (named in reference to the zipper-type mechanism), in which the thiol group of a cysteine residue in the N-terminal propart binds to and blocks the active-site Zn atom so that activation occurs when the thiol group is itself blocked, e.g. by *N*-ethylmaleimide or a heavy metal, or when the propart is proteolytically excised. (*see also* peptide Velcro)

verdi-: A prefix derived from the French for green.

very-long-chain fatty acid: (*see* medium-chain fatty acid)

vesicle: A membrane-limited sac in the cytoplasm of a cell.

vesicular traffic: The transport of intermediates and finished products from intracellular membrane sites to other intracellular or plasma membrane sites.

VH: The variable domain of an immunoglobulin heavy chain. (*see also* V_L)

virion: The extracellular form of a virus, complete with its capsid.

viroid: A nucleic acid that is infective in plants. Unlike a true virus, it has no associated protein

virus: A primitive life form; a nucleic acid surrounded by a protein coat that requires for its own propagation the infection of a plant or animal cell or a bacterium; viruses include important pathogens and useful agents for the study of cellular processes.

visual cascade: The sequence of enzymic and non-enzymic events that is triggered by absorption of light in the retina and results in transmission of a nerve impulse.

vita-: A prefix derived from the Latin for life.

vitamin: A compound required in the diet in minute amounts; fat-soluble if soluble in organic solvents (vitamins A, D and E); water-soluble if soluble in aqueous solutions (ascorbic acid, biotin, cobalamin, folic acid, nicotinamide, pantothenic acid, pyridoxine, riboflavin and thiamin).

VL: The variable domain of an immunoglobulin light chain. (*see also* V_H)

Vmax: (= maximum velocity (V_{max}))

VNTR: (= variable number tandem repeat (VNTR))

VOA: (= vibrational optical activity (VOA))

W

Walden inversion: A change of the configuration at an asymmetrical carbon atom due to the entry from one side of the centre of an attacking group, simultaneously with the departure from the other side of a leaving group.

Warburg apparatus: (*see* manometry)

Warburg-Dickens pathway: (= pentose phosphate pathway)

Watson strand: (*see* Crick strand)

Watson-Crick base pairs: The base pairs that are compatible with a DNA double helix; i.e. adenine with thymine, and guanine with cytosine.

Watson-Crick model: (= double helix (Watson-Crick model))

wax: The product of esterification of a fatty acid with an alcohol other than glycerol, usually a long-chain alcohol.

weaning: The removal of a young mammal from a diet of its mother's milk.

weanling: A newly weaned animal.

western blotting (immunoblotting): A technique to detect specific proteins. A mixture of proteins is separated by polyacrylamide-gel electrophoresis, blotted on to a plastic sheet (e.g. of nitrocellulose) and then exposed to a radiolabelled immunoglobulin directed against the desired protein, which is revealed by autoradiography. (*see also* electroblotting; northern blotting; Southern blotting; south-western blotting)

white muscle (fast-twitch muscle): A poorly vascularized form of skeletal muscle characterized by little myoglobin and few mitochondria; supplied with energy mainly by glycolysis. (*see also* red muscle (slow-twitch muscle))

wild type: In genetics, the unmutated allele or the natural allele that predominates over other variants in a population.

wobble: The presumed slackness of structural requirements for complementarily in fitting codon to anticodon that permits several bases in the third position of the codon to pair with a particular base in the first position of the anticodon.

Woesean tree: A dendrogram that shows the presumed evolutionary divergence of the three dominions, Bacteria, Archaea and Eukarya from a common ancestor, the Last Universal Common Ancestor (Luca).

Wood-Werkman reaction: The proposed reaction, derived from the observed incorporation of isotopic CO_2 into succinate, by which CO_2 condenses with the 3-carbon compound pyruvate to form a 4-carbon compound, oxaloacetate. The basis of the observation was later recognized as the pyruvate carboxylase reaction.

writhe: (*see* node)

writhing number (W): A topological property of double-stranded DNA; the number of times one double-stranded segment crosses another in circular DNA, commonly referred to as a supercoil; equal to the linking number minus the twisting number ($W=L-T$). (*see also* node)

X

X-ray crystallography: A method for structural analysis of solids that have a repeating unit; dependent upon the ability of electrons to scatter X-rays, and capable of locating in space atoms larger than hydrogen in proteins and nucleic acids as well as in less complex compounds.

xaptonuon: (*see* junk DNA)

xeno-: A prefix derived from the Greek for foreign, which indicates origin outside the organism.

xeno-oestrogen: (= endocrine disrupter)

xenobiotic: A non-biological compound, often one that an organism must eliminate or neutralize by some detoxification strategy.

xenoduplex analysis: A method of comparative gene mapping, in which PCR fragments from a species in which the gene sequence is known, forms an interspecies hybrid, a xenoduplex, with a PCR fragment from a species in which the gene has not been identified. The presence of the xenoduplex is detected by its size, therefore its electrophoretic mobility, being somewhat different from that of either of the homoduplices. The method relies on the sequences of the two species being sufficiently similar so that primers designed for the sequence of one species' gene will hybridize to the template of the other species, and that the xenoduplex will be sufficiently stable to be formed and survive electrophoretic separation. When a panel of narrowly selected fragments of chromosomes from one species reacts positively by this test with a recognized gene fragment from another species, the assignment of the gene can be made to the chromosome region from which the panel was taken.

xenograft: A solid patch of foreign tissue that has been grafted onto a test animal for the evaluation of its immunogenicity or carcinogenicity.

Xenopus oocyte: An egg of the African clawed frog, *Xenopus laevis*; a favoured heterologous expression system that, because of its large size, can be microinjected with cDNA or mRNA and monitored for synthesis of a specific protein.

Y-DNA: (= Y-joint)

Y-joint: A DNA heteroduplex structure in which one strand is extended by a non-homologous internal sequence. The loop of the longer strand is base-paired with itself in a hairpin structure. In an *open Y-joint* the shorter, continuous strand is nicked at the joint; in a *closed Y-joint* the continuous strand is un-nicked.

Y-junction: (= Y-joint)

Y-structure: (= Y-joint)

y-type ion: (*see* a-type ion)

YAC: (= yeast artificial chromosome (YAC))

yacto-: A prefix which denotes 10^{-24}; e.g. yactomole, 10^{-24} moles (note, however, this represents less than a single molecule).

yeast artificial chromosome (YAC): A cloning vector with a yeast that can accept a relatively large fragment of foriegn DNA, up to 1 Mb, in yeast cells; a YAC has telomers on each end and a centromere. Their usefulness is limited by the frequency of extraneous sequences that have been imported into the sequence by homologous recombination. (*see also* bacterial artificial chromosome (BAC); cosmid)

yellow enzyme: One of the two first-discovered flavoenzymes. The 'new' yellow enzyme (das neue gelbe Ferment) catalyses the reduction of molecular oxygen or Methylene Blue by glucose 6-phosphate, as is now understood via reduction by NADPH, and has an FAD prosthetic group. The first-discovered 'old' yellow enzyme has an FMN prosthetic group but has no recognized biological function.

Yin-Yang hypothesis: The now-abandoned theory that bidirectional cellular processes, i.e. those that involve two individual pathways (rather than one that may be 'on' or 'off'), e.g. proliferation/ contact inhibition, glycogen synthesis/breakdown, are regulated by a balance of the opposed actions of cyclic AMP and cyclic GMP.

Z

-zyme: A suffix with the same derivation as zymo-.

Z-DNA: A left-handed variant of the DNA double helix originally observed in a sequence of alternating G and C bases, but possible in sequences of alternating purine and pyrimidine bases. The purines are 'flipped' 1808 about the glycosidic bond to assume the *syn* conformation (as opposed to B-form DNA, in which they assume the *anti* conformation); the pyrimidines remain in the *anti* conformation. (*see also* A-DNA; B-DNA; P-DNA)

Z-scheme: In the Hill reaction of photosynthesis, the flow of electrons from water through photosystems II and I to reduce NAD^+.

zebrafish : *Danio rerio*. A model for genetic embryological analyses, which is unique among vertebrates in shortness of generation time, number of progeny and relative ease of examination of embryos.

Zeeman splitting: The detail of a spectrum (nuclear magnetic resonance, magnetic circular dichroism, etc.) that is due to the energy differences of spin states in a magnetic field. (*see also* hyperfine splitting)

zepto- (z): A prefix which denotes 10^{-21}; e.g. zeptomolar, 10^{-21} M.

zero-order kinetics: The insensitivity of a rate on the concentration of a reactant; especially for an enzymic reaction at substrate concentrations much above the K_m. (*see also* first-order kinetics)

zeugmatography (spin imaging): Nuclear magnetic resonance imaging of the distribution of a specific compound, e.g. $^{31}PO_4$ species, in a part of the body, for research or medical diagnostics.

zinc finger (metal-binding finger): A motif of which seven or more may appear in a DNA-binding protein. Each is characterized by two closely spaced cysteine and two histidine residues that serve as ligands for a single Zn^{2+}. When bound to Zn^{2+} they form a module from which protrude amino acid side chains that interact with the bases partially exposed to the DNA major groove. Since in the area of contact there is a limited and local parallel of DNA and protein sequences, the occurrence of the contacting amino acid residues in the protein sequence falls into a regularity, or register, that creates in the protein a syllabicity.

zinc proteinase: A metalloproteinase that features a Zn^{2+}, e.g. carboxypeptidase A, mammalian collagenase.

zipper interaction: The sequence of events by which an immunoglobulin G-coated lymphocyte draws the macrophage plasma membrane around it. Fc receptors on the membrane surface attach the immunoglobulin G markers and eventually surround the cell and engulf it in a phagosome.

zipper-type mechanism: A mechanism for proteolytic degradation of a native protein, in which cleavage of the first peptide bond in each molecule is much faster than subsequent cleavages. The result is that every native molecule has one bond cleaved before any molecule is further affected. In the other alternative extreme case, the *one-by-one type mechanism*, initial cleavage of the first bond of the native molecule is relatively slow but subsequent bond cleavages are fast, which results in complete degradation of one substrate molecule before another is attacked.

Znf: (= zinc finger (metal-binding finger))

zona pellucida: A glycoprotein coat that surrounds an oocyte or ovum.

zonal centrifugation: (= density-gradient centrifugation)

zonation: The phenomenon by which specific areas of an organ have characteristic metabolic functions; e.g. alanine aminotransferase activity is relatively high in the cells of the liver served by the portal blood supply (the periportal region) whereas glutamate dehydrogenase activity is relatively high in the cells that feed the hepatic vein (the perivenous region), and glutamine hydrolysis and urea synthesis are prominent in the periportal region while glutamine synthesis is prominent in the perivenous region.

zone electrophoresis: (= free zone electrophoresis; *see* electrophoresis)

zone of adhesion: An area of a lipid bilayer where inner and outer leaflets join, and by which proteins, gangliosides and other products may be targeted to the outer surface of the plasma membrane from their cytoplasmic sites of biosynthesis. Models postulate that such zones are continuous, e.g. they surround a pore through the membrane, or occur stepwise via a vesicle formed by the pinching off into the inter-leaflet space of part of the membrane inner leaflet, which can then fuse with the membrane outer leaflet.

zoo blotting: A technique to assess the degree to which a DNA sequence, usually a coding sequence, is conserved in evolution; a test for hybridization of a probe complementary to the sequence in one species against DNA isolated from cells of a variety of other species.

zoo-FISH: zoo-fluorescence *in situ* hybridization

Zwischenferment: Obselete term for the oxidative pathway that generates NADPH from glucose 6-phosphate, now known to be glucose-6-phosphate dehydrogenase plus 6-phosphogluconate dehydrogenas.

zwitterion: A chemical compound that has positively and negatively charged groups even when it bears no net charge, e.g. glycine, $H_3^+NCH_2COO^-$.

zygo-: A prefix derived from the Greek for yoke, which indicates a pairing.

zygosity: The condition at an allelic site in a diploid genome, homozygosity if it has two identical alleles, or heterozygosity if two different alleles.

zygote: The product of fusion of a male and a female gamete; a fertilized ovum.

zymo-: A prefix derived from the Greek for leaven which originally referred to catalytic activities of yeast cells, then to any biological catalyst.

zymogen granule: A cytoplasmic vesicle that contains a protein prior to secretion.

zymogen: A proenzyme; an inactive precursor of an enzyme, e.g. trypsiongen. Often, the zymogen is proteolytically activated by an endonuclease. In the case of some aspartate proteases, an acid-catalyzed rearrangement allows cleavage and removal of an inhibitory C-terminal sequence, the prosegment.

zymogram: A slab gel that, after electrophoresis, has been stained to reveal the presence of an enzymic activity, e.g. a gel infused or layered with a protein substrate and subsequently treated with a stain for the protein to reveal, as cleared areas, the location of proteolytic activity. (*see also* zymography)

zymography: A histological method for localization of an enzymic activity. The tissue or cell preparation overlays a gel that contains a chromophore- or fluorophore-labelled substrate, e.g. casein for a proteinase assay. After enzymic reaction, and washing or diffusing away of labelled products, the sample is coated with a photographic emulsion, exposed and developed to superimpose the histological sample on cleared areas in the labelled substrate that represent areas of enzymic activity. (*see also* zymogram)

zymosan: Particles of yeast cell walls.

□ □ □